ROAD FORM AND TOWNSCAPE

ROAD FORM AND TOWNSCAPE

Jim McCluskey

THE ARCHITECTURAL PRESS · LONDON

Frontispiece : The Royal Mile, Edinburgh (photograph by Jim McCluskey)

Most of the photographs in the book were taken by the author. Acknowledgements for other photographs appear on page 305.

First published in 1979 by The Architectural Press Limited: London

© Jim McCluskey 1979

British Library Cataloguing in Publication Data
McCluskey, Jim
  Road form and townscape.
  1. Highway planning 2. City planning
  I. Title
  719'.09173'2   TE175
  ISBN 0-85139-548-1

Filmset and printed in Great Britain by
BAS Printers Limited, Over Wallop, Hampshire

ACKNOWLEDGEMENTS
The author would like to express his gratitude to John Medhurst, landscape architect, and to Peter Hunt and Terry Jones, engineers, for reading and commenting on draft chapters of the book. He would also like to thank the Greater London Council and the other authorities and private architects who have kindly allowed him to include photographs, drawings and illustrations of their work. In particular he would like to thank the Department of the Environment for permission to include in Appendix B material from their Design Bulletin 32.

CONTENTS

INTRODUCTION

Roads are not just routes

Roads are not just routes along which people in vehicles move from one part of the environment to another; nor are they just ribbons of asphalt or concrete built to accommodate vehicles, from which the outside environment can be observed at the windows like the flickering images of a television screen. To a large extent the road with its setting of buildings or landscape *is* the outside environment. There are still people who work in fields and we have our parks, landscaped open spaces, nature reserves, hill walks and so on. Nevertheless for most of us most of the time the public outside environment is the road or the street and its surroundings.

All roads are not designed to fulfil the same function. Although they are all initially conceived as routes, ways for travelling from one place to another, they also have the properties, to some extent, of a place, a location in which people enact part of their lives. The degree to which a road is a place or a route varies. A village road, where it widens out to accommodate the stalls for the local market is predominantly a place, whereas the inter-city motorway designed as it is for high speeds and efficiency in proceeding from one place to another is predominantly a route. Yet even the motorway can be conceived as a series of places. We tend to remember a journey along a motorway in terms of interesting lengths and dull lengths, stretches with good views, others which were contained by the sides of a cutting and yet others which were elevated and of which the main impression was that of sky and wide vistas.

Unity of design

When a person describes a stretch of road, therefore, it is in terms of the way he experiences the whole environment. The road is not perceived as just the surface of concrete or asphalt. As well as being the surface it is the trees and shrubs at the side of the paved area, it is the views seen through the gaps in the trees, it is the sides of the cutting and the containing hedgerows. In the urban context it is also the buildings which line the route, the bridge overhead, the trees lining the avenue, the landscaped area in the square.

All these aspects of the surroundings come together in a person's perception and the road is seen as a composition of all the elements in view. The road can be thought of as analagous to a room or a series of rooms: the surface is the floor, the walls are the buildings, the vegetation, the surrounding hills or the open view, the ceiling is the surface implied by the tops of the buildings on either side and the canopy of trees or the sky.

The person using the road may experience his surroundings at a particular time as a complete unit, as a number of units linked in series, as a unit which is constantly changing or as some combination of these three. Although the user experiences his surroundings in these ways, those involved in the design process have not always viewed in a similar manner the environment which they were creating.

In the design of any road many professional disciplines are involved. There is a tendency for the various aspects of a design to be allocated to different professionals. These professionals may not have the same objectives and will certainly have different training from each other. In a particular residential development the planner may be

mainly concerned with such matters as density and zoning, the architect with house types and the groupings of dwellings, the highway engineer with the efficiency of vehicle access, the quantity surveyor with overall cost and the services engineer with the economic routing of services. Their specialization and training can be such that it is difficult for the members of one profession to participate in the aims and problems of the others. It has become apparent that this system of specialization has sometimes contributed to serious deficiencies in our surroundings and that there is an urgent need for all those involved in the design process to be conscious, at all stages, of the total end product and not just their own part of it.

So this way of conceiving a road as a three-dimensional entity means that a decision arrived at within a particular speciality should be reached in the light of those which are being made by the members of other professions working on the design. In a residential area the decisions regarding the colour and texture of the road surface should be made in the knowledge of the types of facing brick to be used in the construction of the dwellings and should not only be based on low cost and standard practice; the choice of lamp standards should be a function of the forms the buildings are to take and not simply those which are most easily available; the kerb details should be dependent on whether the development is to have an urban or an arcadian character and not used merely because a standard detail prescribes it, and so on.

An understanding of other professions

In order to adequately overcome the difficulties inherent in the need for roads to be designed as total entities when the designs are prepared by teams of specialists, it is necessary that members of each profession acquire some understanding of the other disciplines involved. It is therefore desirable that publications be made available which will be of interest to a wide spectrum of the professions involved in the design process.

There is also a need for publications which will be of interest and comprehensible to laymen who have a special responsibility for the quality of the environment: politicians, local authority councillors, developers, members of local amenity and environmental groups. These have been borne in mind in the preparation of this book.

In particular the difference in training between architects and highway engineers can result in conflict and misunderstanding. It is for this reason especially that architectural and engineering aspects of design have been included in one volume. If an architect is designing a housing area it is important that he is aware for example of the structure of the asphaltic pavements. This will better enable him to choose with the engineer a wearing course whose colour and texture will best integrate the road into the total design. If an engineer is designing a road in such an area it is important that he understands the principles of the containment of space so that he can work with the architect in creating a unique place. In the design of a major rural road the landscape architect should be knowledgeable about the principles of flowing alignment just as the engineer should have an understanding of how the road alignment can be chosen to enhance the landscape, how planting can be used to open and close views and so on.

Obviously no-one can master all the professions involved in such undertakings but in these areas some understanding of the other disciplines is essential if the partnership is to be fruitful.

Characteristics of different road types

Before a road can be designed in this way, as a total entity, it is necessary to establish the desirable characteristics of a design. The reasons for building any road will involve the satisfaction of needs of individuals and groups of individuals in the community. These needs should be clearly established at the outset and agreed by all those involved in the project. Failure to do this may well result in the construction of a road which will not fulfil its proper purpose. There has been a tendency in the recent past to design roads as though high speeds were always desirable and as though they should always be built primarily for the use of motor vehicles. This is not necessarily the case. Such a view would not be shared by a person who enjoys wandering through an English village, driving along a tortuous route in the Scottish Highlands, walking up the Royal Mile in Edinburgh or round a crescent in Bath. Nor would it be shared by a person who knows that his children are safe because the vehicles in his housing area will not be travelling at more than 10 miles per hour.

Some characteristics are common to all roads. All roads must be safe, all roads should be an integral part of the total environment, enhancing it rather than detracting from it and all roads should be enjoyable to use. That they all must be safe is accepted and the reasons for integration and unity have been mentioned. The need for roads to be enjoyable places as well as routes is a consequence of the fact that they constitute a large part of our outdoor living space. They should be designed in such a way as to give maximum satisfaction to those using them on foot, on bicycles and in vehicles.

The concept that roads should give pleasure to their users is one which has probably been afforded more conscious consideration in the design of rural motorways than most other types of road. In the design of these motorways great consideration is commonly given to the merging of the road alignment into the land form, to the appearance of overbridges, the integration of cuttings and embankments into the landscape and other matters relating to the appearance of the total road environment. Such concern about the pleasure-giving aspects of designs has not always been shown with other road types. In particular it is often lacking in the design of roads in residential areas, where the tendency has been to concentrate on the provision of generous areas of running surface for vehicles and on the standardization of layout geometry and materials, pedestrians usually being relegated to a surfaced strip on either side of the road pavement.

There are also characteristics which are common to a group of roads but not to all roads. Some will be built to accommodate high speeds and the alignment will include transition curves and large radii; some will be built to cause minimum disruption of a beautiful landscape and may be narrow, serpentine and have edges which merge into the landscape; some will be built to be used by vehicles and pedestrians and will be designed to ensure slow vehicle speeds.

When roads are considered in this way, it is possible to group them in categories determined by the dominant characteristics. I wish to propose four categories, each of which is of major importance. They have been named in terms of alignment and are:

the townscape alignment
the flowing alignment
the hillroad alignment and
the countryside alignment.

If we understand the proper attributes of each alignment type, we will have taken a significant step towards the creation of roads which will be in sympathy with their surroundings and which people will enjoy using.

The townscape alignment is that appropriate for use in areas where the needs of the pedestrian are dominant and vehicles are constrained to travel at low speeds. Thus it encompasses the design of roads in hamlets, villages, towns, neighbourhoods within cities and the suburbs of towns and cities.

In other words, it includes all urban roads which are not distributor or trunk roads. A townscape alignment is one which can be enjoyed at walking pace; it creates an environment which is sufficiently integrated, varied and interesting to give pleasure to pedestrians using it at the same time as it allows vehicles to circulate at a pace which does not disrupt that environment.

The flowing alignment is that which is designed to accommodate vehicles moving at high speed through a landscape. This alignment is properly designed in such a way as to enhance the countryside through which it is constructed and at the same time to provide an enjoyable and stimulating environment for the occupiers of the vehicles. The principles of flowing alignment are well understood and reasonably well documented, notably in *Man-made America* by Tunnard and Pushkaref.[1]* A well-designed alignment of this type satisfies the demands of man, machine and landscape. The long sweeping three-dimensional curves which result from a proper combination of vertical and horizontal alignment are a delight to the eye as well as perfectly accommodating the dynamic demands of a vehicle in motion. The elegant curves hug the contours and elucidate the land forms. The driver enjoys the kinetic sensations and the slowly unfolding view.

The hillroad alignment is the type which is indispensable if we are to preserve the character of our highland areas. Typical of this category are the roads found in the Scottish Highlands. Narrow and serpentine, the road clings to the contours of the glens and hillsides. The traveller on such a road experiences *being in* the landscape. He is constantly changing direction and elevation in sympathy with the land forms. The scale of the road is such that he is not alienated from his surroundings and his sense of being in the landscape is heightened by the vegetation coming to the edge of the running surface. If such roads are superseded by flowing alignments, the intimate relation between the driver and his surroundings will be completely destroyed. He will no longer be able to experience *being there* as is possible at the present time. The views will flash past like pretty pictures on a screen and the tourist will move from his centrally heated car to his centrally heated

hotel in Glasgow, Inverness or wherever.

The periodic straightening of corners and widening of straights is already beginning the process of altering the character of these roads. A detailed analysis of this type is urgently required since only a thorough understanding of their characteristics and the unique experience of the landscape which they make possible will enable unsympathetic 'improvements' to be resisted.

The countryside alignment is the lowland equivalent of the hillroad type just described. Its scale is much smaller than the flowing alignment. It follows much more closely the local contours and will move aside to avoid a farm steading or dip into a depression which a major road would glide over on embankment, yet it is not as serpentine or dramatic as the hillroad. It shares with the hillroad an absence of conspicuous edge detailing and has a similar intimate relationship to the surrounding vegetation. In some areas of south-east England the cornfields come right up to the road edge on either side; elsewhere the road negotiates field boundaries and is contained by hedge or walling.

Here again the underlying principles governing the design of these roads require careful study if they are to be maintained. And it is important that they should be maintained for the same reasons that we must protect the hillroad alignments; not only does their form determine the quality of the users' experience, but it also greatly affects the quality of the landscape as seen from outside the road.

An approach to road design which embodies the four types outlined above would provide a framework within which highway engineers could direct their attention to the solution of some of the environmental problems which are now of such urgent concern to society. It would enable us to approach the problem of integrating our roads into the environment armed with a method of tackling road design from the point of view of the way people experience their outside living spaces. Such an approach will avoid the purely technical/geometrical solutions which can so dehumanize our surroundings, and may lead us to a method of dealing with the apparently intractable difficulty of coping with the car in our cities without comprehensive redevelopment on a massive scale; it should also assist in sorting out the proper priorities involved in providing tourist access to the countryside.

A classification has purposely not been made for urban motorways. This would apparently come under the heading of flowing alignment yet it is doubtful if a viable solution for this problem has yet been discovered for British cities, and to raise this issue here would lead us outside the scope of this book.

This volume is primarily concerned with two of the categories listed, the townscape alignment and the flowing alignment.

There are many levels at which townscape alignment can be studied. Here attention has been focused on presenting to the reader part of an alphabet which can be used to create a language of urban environment as it relates to roads. This alphabet will be added to as new designs are developed and as greater understanding is gained of the well-loved parts of our existing surroundings.

It is imperative that the ingredients of good townscape are used in a meaningful way. Clearly there is a need to instil visual interest, variety and order, but the form of our

urban surroundings can have more profound symbolic meanings which are being investigated by a number of writers. A competent designer will be familiar with the many aspects of the subject at all levels.

A large part of this book relates to minor rather than major roads. Concern is often expressed today about the deterioration in the quality of the environment especially in those areas where people have their homes. One of the keys to understanding how we may control environmental quality lies in the design of roads. It is accepted in many quarters that, in the recent past, the premises on which the design of minor roads in built-up areas was based were not always the most appropriate. It is partly because of confusion regarding the proper design premises for local roads as opposed to major ones that the approach to local road design has come under criticism. Thus the types of design named have been selected for examination because a major concern of this volume is a clarification of an appropriate approach to the design of minor roads as opposed to that which is proper for major roads.

Summary

To summarize, the first main concept which underlies the contents of this book is the need, during the design process, to integrate all aspects of road design so that the end product will be a coherent whole, a unique part of the outside environment which people will enjoy. This requires a good understanding of the various disciplines involved. Secondly it is necessary that designers clearly comprehend the desirable attributes of roads depending on their purpose. Detailed consideration is given to townscape and flowing alignments.

Two aspects of road design which must be considered with both types of alignment are the influence of land form, including road earthworks, on the design and the use of planting as an integral part of a scheme. A chapter is devoted to each of these subjects.

The proper use of materials from a visual point of view is considered in Part IV, Materials, Surface and Trim. This is mainly relevant to the townscape alignment since with high-speed roads the engineering properties of surface materials in relation, particularly, to skid resistance and durability will be strong determinants of the materials used and thus of their colour and texture. Nevertheless even here considerable control of texture and colour may be possible in the choice between rigid and flexible pavements, the use of colour additives to asphalt and concrete, and the choice of aggregates.

The configurations of flowing alignment are largely determined by the dynamics of vehicles moving at high speeds. Appendix A has been included to illustrate how this is so by showing the relationship between the size of road curves and the speeds at which the vehicles move. On the other hand the slow design speeds which are now being recommended for residential areas require the application of special design data and Appendix B relates to this data. The presentation of the material has been made as visual as possible by the extensive use of photographs and diagrams. This is appropriate for any type of subject matter, since a large amount of environmental information can be displayed by one photograph and they have the capacity to convey the overall character of a place. They illustrate whether or not a synthesis of all the elements of a design has been achieved and to assist in the achievement of this synthesis is the central purpose of this book.

PART ONE

THE TOWNSCAPE ALIGNMENT

People move well on their feet. This primitive means of getting around will, on closer analysis, appear quite effective when compared with the lot of people in modern cities ...
People on their feet are more or less equal. People solely dependent on their feet move on the spur of the moment, at three to four miles per hour, in any direction and to any place from which they are not legally or physically barred. An improvement on this native degree of mobility by new transport technology should be expected to safeguard these values ...
Ivan Illich, Energy and Equity

1
NETWORKS

A system of access roads constitutes a network which makes possible the movement of vehicles and people through the built environment. Different networks have different characteristics. An examination of these differences will assist the designer in deciding on the most useful type of network for a given set of circumstances. If a housing area consisting of a number of dwelling clusters is to be serviced then a branching network may be required, whereas if a primary consideration is direct interconnection between local areas then some form of grid network may be indicated.

In this chapter we will examine the important diversity of choice made possible by the characteristics of different network types. We will then illustrate with design examples how this diversity of choice can be and is being used.

In order to establish the context for this examination of networks it is necessary to consider briefly the way in which we conceptualize our environment.

Places and paths

It is of primary importance to man to be able to orient himself in his surroundings. He must retain and organize his knowledge of his environment in order to be able to act effectively within it.

The world of the newly born baby is limited to his immediate surroundings, to that which he can touch, taste and smell. Later it is extended to the limits of his vision. As he becomes able to move from one place to another, as his faculties develop and as he becomes more experienced, his world is extended to other areas than those which he can actually see. To assist in the task of retaining and organizing his experience of the environment the individual constructs for himself a mental image of his world. This mental image has certain basic elements common to all of us. Three of the most fundamental elements have been called place, path and domain.

A place is a contained area which is known and to which we attribute a certain character. Places are areas in which significant events of our life occur. The concept of place implies an inside and an outside. When we go outside a place we are setting off on a path or route to another place.

Paths are the routes connecting places. Whereas a primary attribute of a place is its quality of being contained, even if it is only by an imaginary boundary, the linear quality is primary in the case of path. Paths imply movement. Places imply pause or rest.

The domain is the ground in which places and paths have their existence. The image of the domain is less well defined than that of the path, and for the purpose of our present study the place and the path are the important images.

In the diagrams used to illustrate network patterns we will represent a place by a circle O and a path by a line ———.

Network patterns

In order to simplify our examination of patterns it will be assumed that each network consists of a number of places connected by routes. This is not strictly the case when buildings have direct access to routes along their length. Then the route itself takes on some of the characteristics of a place. This complication will be considered later and in the meantime the simplification in our assumption will

clarify the discussion without invalidating our findings.

We should note at this point that this examination of networks is presented in order to assist in the understanding of some interesting aspects of network patterns in relation to roads. It is not presented with a view to proving that one network type is cheaper or better than another. There are a great number of factors which must be taken into account in making a choice. The characteristics of a particular network, as described here, is only one of these factors. (The attributes of network patterns are considered in some detail in *Patterns in Nature* by R. L. Stevens.)

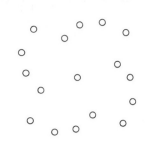

1.1

SERIAL AND RADIAL PATTERNS

We will now look at two, fundamentally different ways in which a number of places can be joined by routes. An arrangement of places around a central place is shown in **1.1**. The particular choice of positions for the places has been made to ensure clear and simple diagrams. The points being made are equally valid for random dispositions. For ease of reference we will call the central place the origin and we will look, to start with, at two ways of joining the origin to the other places.

One way is to join it to one place and then to join that place to a subsequent place and so on, until all the places are linked together like beads on a string. This is the arrangement of **1.2**. We have called it the serial pattern.

On the other hand each place can be connected to the origin directly by means of its own unique route. This solution is shown in **1.3** and has been called the radial pattern.

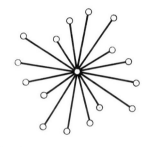

1.2

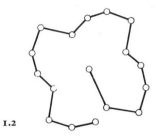

1.3

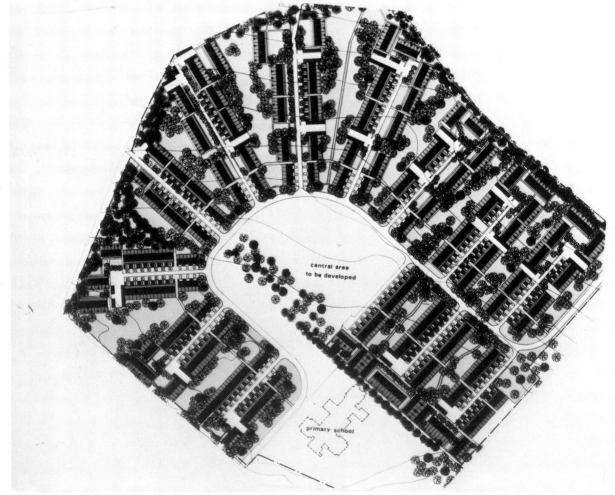

central area
to be developed

primary school

1.4

The characteristics of these two types show noteworthy contrasts. The total length of the path is minimized with the serial pattern and, in the case illustrated, can be represented by 10 units for the purpose of making comparisons. On the other hand if we consider the total length of routes from the origin to each place in turn and then take the average, this figure is 5·5 units which is high when compared with the average length of routes for other types of network.

In the case of the radial pattern the total length of path is 20 units. This is twice as long as with the serial pattern. But the average length of path from a place to the origin is only 1·2 units, approximately one fifth of the average path length with the previous arrangement.

The difference in values quoted for the total route length and the average route length from the origin are not simply characteristic of the example chosen. On the contrary they indicate attributes of these two patterns whatever the particular location of places.

The serial pattern provides a highly indirect method of connecting a number of places to one origin whereas the radial pattern enables the places to be connected to the origin in the most direct way possible.

With the serial arrangement each route from a place to the origin passes through all the intervening places. On the other hand with the radial pattern there are no intervening places.

The serial pattern affords the shortest total length of route whereas the radial pattern results in the longest total length of route when connecting an origin to a number of places. It is interesting to note that in terms of access roads of constant width, everything else being equal, the serial mode would minimize capital cost (since the road length is a minimum) and maximize journey costs (since average length of journey is a maximum). The radial pattern would maximize capital costs and minimize journey costs.

An advantage of the serial pattern is that no more than two paths connect into one place, whereas the radial arrangement necessitates a multiple connection at the origin which could present difficulties in practice. These difficulties can be overcome, and one method is illustrated in the access road network of **1.4**.

The serial system indicates a desirable arrangement of roads when it is required that the path should lead through each place in turn as could be the case with, say, the route of delivery or other service vehicles.

The radial pattern may be indicated when there is a very strong focal area to which all paths should lead; for example to a town centre or a shopping and recreational area.

In residential areas the equivalent of a place might be a cluster of dwellings and the paths would be the roads connecting the clusters. In practice the distinction between paths and places is often not as obvious as this.

It is of interest to consider a simple modification of the serial pattern. If the serial mode of connection is maintained but the paths link with a path running transversely through the origin, as shown in **1.9**, then the route characteristics alter considerably. The total path length becomes 12 units, an increase of 20%, but the average path length from a place to the origin becomes 2.5, a reduction of 55%.

For example in the case of dwellings arranged along a road it is possible to view each individual dwelling as a place which is linked in series to the other dwellings by the road, **1.5**. In the case of a radial network of roads in which dwellings were aligned along the road the system would combine the characteristics of the radial pattern with those of the serial pattern.

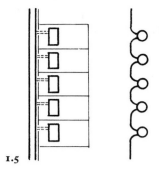

1.5

BLOCKAGES
The serial pattern is very sensitive to blockage of a particular path. If one link between places is blocked all the routes connecting places on opposite sides of the blockage become ineffective, **1.6**. This is not so with the radial pattern. In this case a blockage will only interfere with the path between one outlying place and the origin, **1.7**.

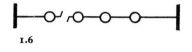

1.6

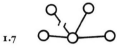

1.7

One way of overcoming this disadvantage of the serial system is to connect two serial patterns in parallel. Then when one series is broken the other can be used, **1.8**.

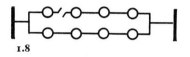

1.8

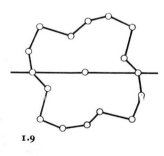

1.9

This example is analogous to the loop road used in access road networks. The use of the loop road is common in the design of residential layouts, **1.10**.

THE WEB PATTERN
The characteristic of the simple radial pattern—that all paths between places outside the origin must pass through the origin—can result in very circuitous routes. A modification which improves its performance in this respect is shown in **1.11**. This will be called the web pattern.

When this arrangement is put into practice the areas bounded by the paths can take on awkward shapes, posing difficult problems for the designer.

Nevertheless the pattern, either in its pure form or modified, is sometimes used for road networks. An example of the former is that of a late 16th-century attempt to build the ideal city on a site some distance outside Venice, **1.12**.

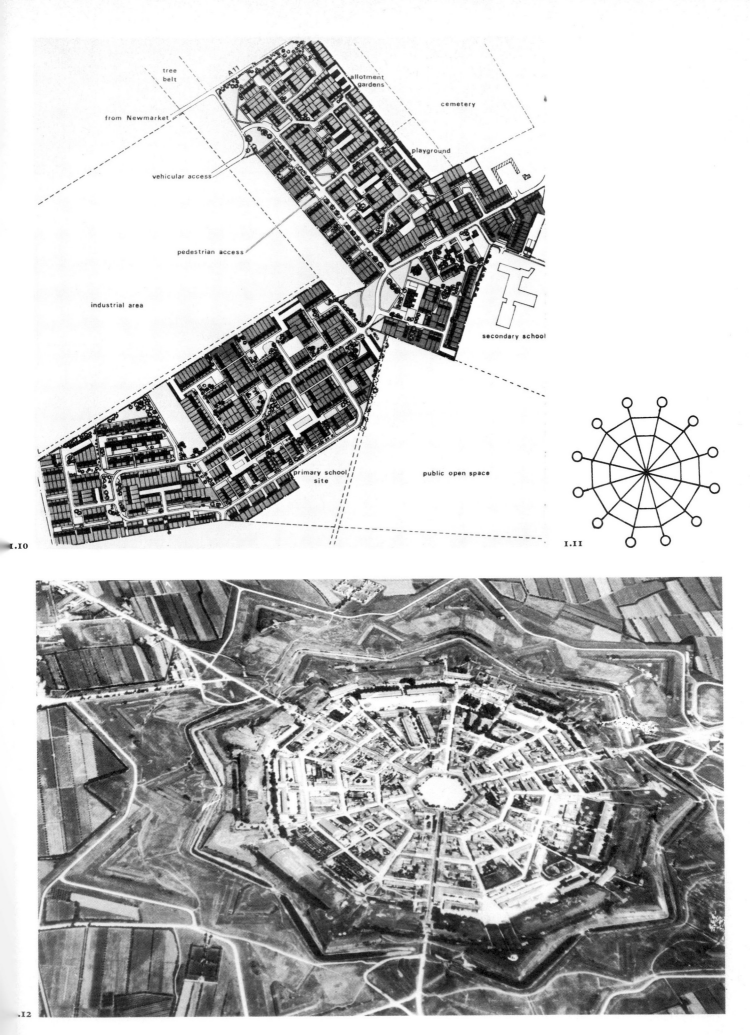

tree
belt

A11

from Newmarket

vehicular access

pedestrian access

allotment
gardens

cemetery

playground

secondary school

industrial area

primary school
site

public open space

1.10

1.11

.12

15

THE BRANCHING PATTERN

Another way of modifying the radial pattern, **1.13a**, is to dispense with some of the direct connections between place and origin and to connect these places to the remaining radial paths. When this is done, **1.13b**, the resulting total path length is 15.3, a reduction of 24%, whereas the average path length becomes 1.4, an increase of 17% on the purely radial case. This modification can be considered as intermediate between the radial pattern and the branching pattern.

For the branching pattern shown in **1.13c** the total path length is 12.7, a reduction of 33% over the arrangement of **1.13a**, and the average path length is 1.7, an increase of 42%. The branching pattern has the added advantage that all the junctions can be made suitably simple. Also the paths can be evenly distributed over a given area by choosing an appropriate arrangement of branching elements. This means that the areas between the paths can be made similar in size and shape, **1.14**. Also the shapes can be easily designed to be those which best suit their purpose. This is much less easy with the radial pattern.

A branching system produces a series of dead ends, inhibiting free-flowing movement through the area. For an access road network this may or may not be a desirable feature depending on the type of development being designed. It is often chosen when the wish for privacy and quiet outweighs other considerations.

An important feature of this system is that it is composed of a series of elements which can be treated as being arranged in a hierarchical order. This characteristic can be used to attain economies by designing each element to fulfil its particular function in the hierarchy.

It is evident that this property is significant in the design of road networks since with a branching layout there is a hierarchy of traffic flows and it can be advantageous to have the option of designing particular elements for particular flows. However tailoring the size of the road element to suit local conditions within a system is also a relatively simple matter with the serial system as it is with the grid network.

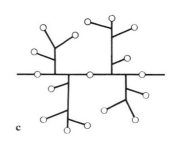

1.13a

b

c

1.14

THE GRID NETWORK

If we now consider a regular branching pattern as depicted in **1.15a** we can see that by joining up the branches the resulting network takes the form of a two-directional grid, **1.15b**.

By altering the branching pattern in this way the 'dead-end' element has been eliminated. The new network allows movement through the area from one point on the perimeter to another. It affords a large choice of starting and finishing points, and it provides a large choice of routes between any pair of these points. The directness of the routes which it is possible to take will be a function of the 'grain' of the mesh established. A grid arrangement of paths thus provides greater freedom of movement and offers more direct routes between places than is possible with a branching pattern.

Many other types of grid network can be developed. Figure **1.16** shows a three-directional grid. For the same grain a three-directional grid provides more direct paths than a two-directional one.

A serious disadvantage of a regular grid of this type in the context of access road design is the awkward triangular areas which are formed between the paths. Another disadvantage, which it shares with the regular rectangular grid, is the junction of a large number of paths at the same node. This is not necessarily a problem if the junction occurs at a place which covers a sufficiently large area, but if it is merely the meeting of a number of routes as at a cross-road then this can create an unsatisfactory feature. In fact for this type of grid the disadvantages outweigh the advantages to such an extent that it is never used in its pure form.

Although the problem of four paths meeting at one node occurs in the branching pattern, **1.15a**, it can be easily overcome by ensuring that branch junctions are made alternate rather than opposite, **1.17**.

A similar solution is possible with a rectangular grid, **1.18**. A grid need not be made regular and there are many reasons why a designer may decide to introduce irregularities. For example he may wish to curve the paths for aesthetic reasons, or the existing ground contours may dictate a curved form, **1.19**.

The plan of part of the layout for Welwyn Garden City, Hertfordshire, designed in accordance with the Arcadian precepts put forward by Ebenezer Howard,[2] provides an example of the use of an irregular grid, **1.20**.

A grid network is easily adapted to accommodate changing patterns of flow. A possible system of one-way flows through a rectangular system is shown in **1.21**.

A regular grid pattern has the useful property of providing equal coverage over its whole area. At the same time the grain is easily altered to provide for changing conditions, **1.22**.

In access road layouts a grid is often used to provide a large-scale mesh while the units which it contains are covered by some other type of pattern. In the example shown in **1.23** the areas enclosed by a rectangular grid pattern are covered by a branching pattern.

The grid pattern is not sensitive to interruptions in particular elements. This is obviously of some advantage in an access road system if lengths are subject to closure due to such activities as the repair of services and resurfacing.

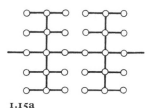
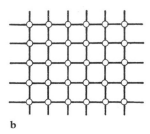

1.15a   b

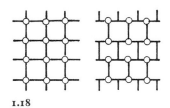

1.16

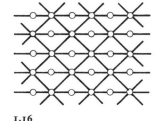

1.17

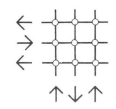

1.18

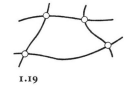

1.19

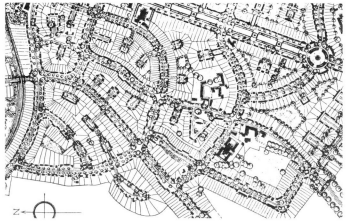

1.20

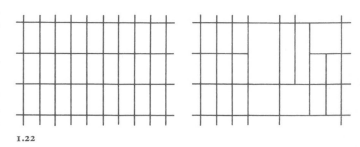

1.21

1.22

17

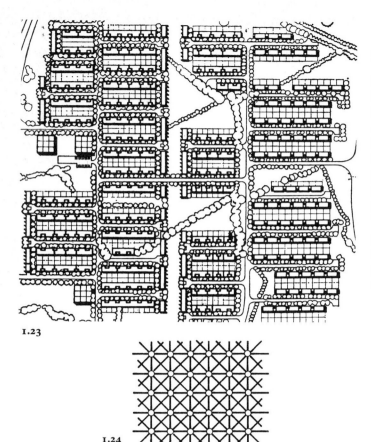

1.23

1.25

3L. Otherwise it will be necessary to provide a network for service vehicles and the same options are open for these as for access roads generally.

Although it has been normal practice up to the present time to provide access roads with two lanes (one for each direction of traffic) in certain circumstances it is feasible to provide one lane only. This single lane may support two-directional traffic or, possibly, in the case of loop roads and through roads, one-directional traffic.

Two-lane, one-way loop roads or through roads are also possibilities. In the case of access roads it is unlikely that the volume of traffic will be sufficiently large to justify such a solution. The number of lanes and the direction of traffic will be indicated by the code shown in **1.25**.

1.24

A four way square grid is illustrated in **1.24**. A notable quality of this arrangement is that there is a direct path from each place to all of the eight places immediately surrounding it. This is highly efficient in terms of directness and choice of route. The obvious disadvantages when applied to access roads include the cost, the high land consumption of the multiplicity of routes, the large number of junctions and the unsatisfactory shape of the enclosed areas.

Such an arrangement is never used in its pure form but modifications of this type of grid, which affords a high degree of interconnection, are feasible.

Some examples

The examples which follow are presented with a view to illustrating the great diversity of possible solutions to the problem of coping with the flow of vehicles in a built-up area. All the examples have been taken from residential developments. Of course, those presented by no means exhaust the possibilities. For easy reference a symbol has been drawn for each of the eight types of solution illustrated. These are shown in **1.26**.

Except for cases which are too simple to fall into a network type, a solution will invariably conform to the branching, serial or grid type of network or to some variation or combination of these three.

The first example is a basically simple arrangement in which all the dwellings of the development enclose one space. Such a place could be the node of any of the types of network described. An option which may be open to the designer is the exclusion of private cars from the housing area entirely. This is the second type of solution illustrated. If the site is a narrow one it may also be possible to exclude service vehicles, as in the example of Thamesmead Area

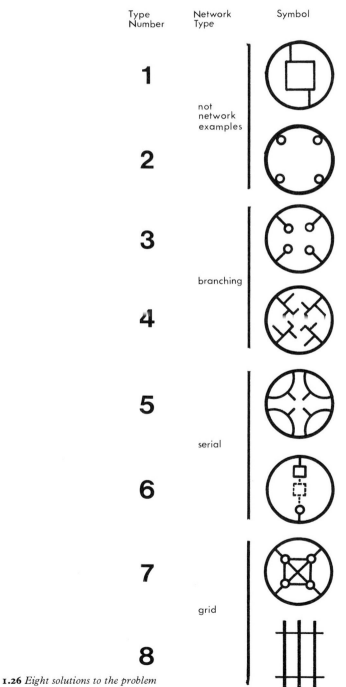

1.26 *Eight solutions to the problem of access in built-up areas*

1 Layout in which dwellings are grouped around a central space.

2 Layout in which all cars are parked adjacent to a perimeter road round the housing area.

3 Layout in which short access roads penetrate the housing area; cars are stored at the end of the culs-de-sac. Pedestrians can pass from the end of the cul-de-sac into the housing area. No other access to the housing area is possible from the access road.

4 Layout in which access is by means of a system of cul-de-sac.

5 Layout in which access is by means of a loop road or series of loop roads. (The loop roads may have associated culs-de-sac.)

6 Layout in which a number of enclosed areas are connected in series.

7 Layout consisting of a grid of interconnected areas.

8 Layout consisting of a grid of interconnecting access roads.

Solutions types 1 to 8 are related, in **1.27**, to the possible number of lanes and directions of flow, A to C. The areas enclosed by the thick lines indicate theoretically possible combinations of 1 to 8 and A to C. The dot ● indicates the combinations illustrated in the examples which follow.

The examples demonstrate the intimate relationship of the choice of network solution to the problems of vehicle access and the character of the residential layout which is being designed. They illustrate that an important aspect of network choice is the need to integrate the road into the overall concept. These studies will provide a suitable background to the examination of particular aspects of road design and townscape in the following chapters.

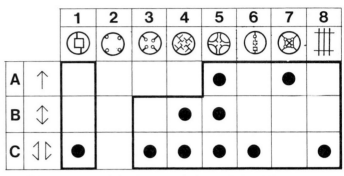

		1	2	3	4	5	6	7	8
A	↑					●		●	
B	↕				●	●			
C	⇅	●		●	●		●		●

1.27 *Diagram summarizing the network characteristics of the examples (below)*

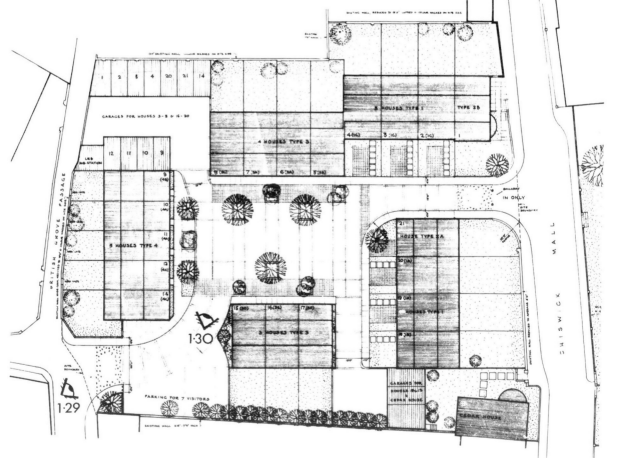

1.28

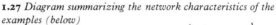

MILLERS COURT, OFF CHISWICK MALL, HAMMERSMITH, LONDON
Dwellings grouped around dual use pedestrian–vehicle court
(Architects: Chapman Taylor Partners)

The architect has chosen a courtyard layout for this highly successful infill development. This choice has enabled her

to design a scheme which is urban in character and embodies a sense of order and peaceful seclusion.

The road is fully integrated into the design and this fact greatly contributes to the sense of unity which is experienced. The designer has treated the road surface with as much careful attention to form, proportion, texture and detail as she has treated the elevations of the buildings. The result is an enclosed area which is both intimate and full of character. In the courtyard cars and pedestrians share the same surface.

I.29

I.30

A view of the north west access is shown in **I.29**. Cars are slowed down on entering the housing area by the rough-textured cobbled surface at the start of the access road, by the raised section of cobbles and by the curvature of the road. The white wall facing alerts the night driver and the road curvature lends privacy to the central dual-use area.

The transverse lines of pavers add a static ingredient to the dynamic line of the road and thereby encourage the driver to further reduce speed, while the broad transverse bands of pavers at the end of the access road imply 'dead slow' to the driver and signal his passing from the linear space of the road to the contained space of the square, **I.30**.

The upstand kerbs used on the access roads have not been extended into the central space. The photograph illustrates the high sense of enclosure and privacy that has been achieved. In spite of the apparent informality of the arrangement of the shrub beds when viewed on location they are nevertheless placed in such a way that the allocated position for each car is clearly defined. An orderly pattern of parking is thus ensured.

The positions of the beds also ensure that the two-way through route for vehicles is clearly indicated, but at the same time they create a route which is sufficiently circuitous to inhibit drivers from exceeding a walking pace, **I.31**.

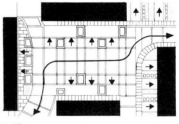

☐ planting beds

I.31

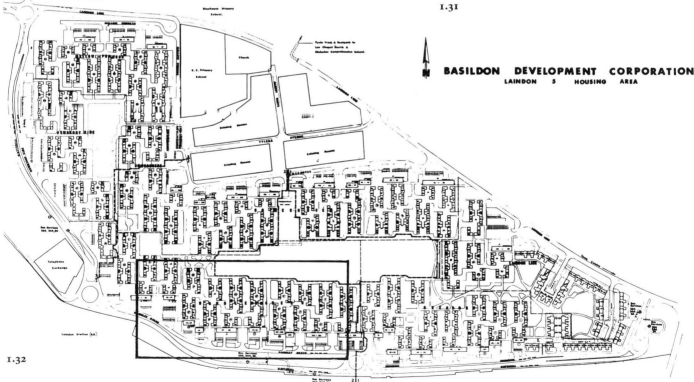

I.32

BASILDON DEVELOPMENT CORPORATION
LAINDON 5 HOUSING AREA

LAINDON 5, BASILDON NEW TOWN, ESSEX
Parking off perimeter road ; no cars within housing area
(Architects: Basildon Development Corporation)

In this arrangement garage courts and areas of hard standings are provided at the perimeter of the development. No private cars are permitted to penetrate the

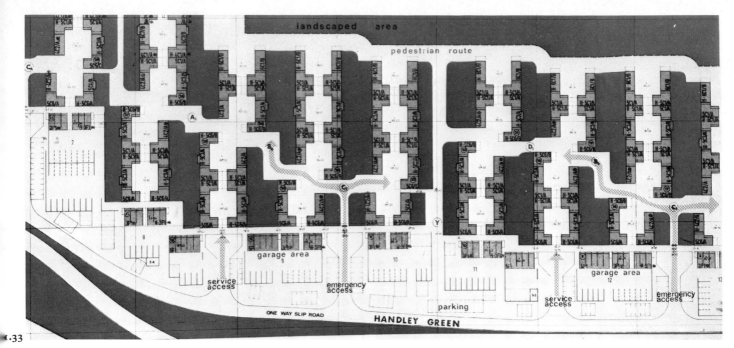

housing area and yet, since the car storage areas are spread evenly around the perimeter, each resident can park within a very short walk of his dwelling.

Access for normal service vehicles is provided as shown in **1.32** and **1.33**. Emergency vehicle access is also shown in the detail plan, **1.33**. These latter routes which are designed to take fire engines and pantechnicons, but are normally the transverse pedestrian ways, have been made sufficiently serpentine to preclude excessive speeds and thus ensure pedestrian safety.

Laindon 5 comprises 1364 dwellings with 125% parking provision, including 788 garages.

The density is 75 persons per acre (ppa) when the central open space is included.

The project architects were Maurice Naunton and John Byron under the Chief Architect/Planner D. Galloway.

1.35

34

This view of the perimeter road, **1.34**, shows an entrance to the walled parking precincts. With a layout of this type it is most unlikely that the driver will park on the road—there would be no point, since he will be more conveniently placed for access to the dwellings by driving into the parking areas.

The small-scale enclosed spaces

of the pedestrian courts (less than 12 m square), **1.35**, provide quiet precincts in which it is easy for neighbours to get to know each other. These precincts are also safe for children at play.

By providing a vehicle-free housing area around a central open space children can safely reach the landscaped area, **1.36**, from their homes.

1.36

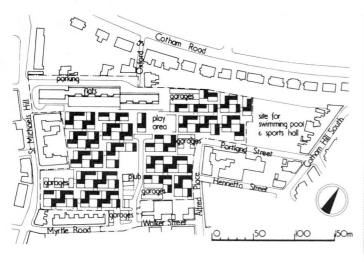

1.37

1.38

HIGH KINGSDOWN, BRISTOL
Parking off perimeter road; no cars within housing area
(Architects: Whicheloe Macfarlane Partnership and
Building Partnership (Bristol) Ltd.)

Although one cul-de-sac has been incorporated in this
scheme it is not essential to the type of layout used. This
example, like the previous one, belongs to that type of
housing layout which does not permit cars to penetrate the
housing area. Instead the residents and visitors park their
cars at the edge of the development and walk to the
dwellings. The locations of car storage areas are marked
'Parking' and 'Garages' on the plan, **1.37**.

The High Kingsdown development includes 103 houses,
110 flats and a supermarket.

Car accommodation is provided on the basis of one closed
garage per unit and visitors' parking spaces are allocated at
the rate of one per three houses.

The design team was composed of Roger Mortimer, Ken
Greaves, Mike Axford, John Haigh and Barry Cannell.

1.38 shows excluded vehicles
parked adjacent to pedestrian-
only access at mid-left of picture.
By keeping cars on the perimeter
of the housing area the architect
was able to design a tightly knit
development penetrated by a
system of pedestrian alleyways.
This arrangement derives from
the system of footways and walled
gardens which were characteristic
of the site before redevelopment.
The sense of enclosure and
privacy which pervades this
scheme, built within walking
distance of the centre of Bristol, is
illustrated in **1.39**.
Some of the lanes have been made
sufficiently wide to take fire
engines and other service
vehicles.

1.39

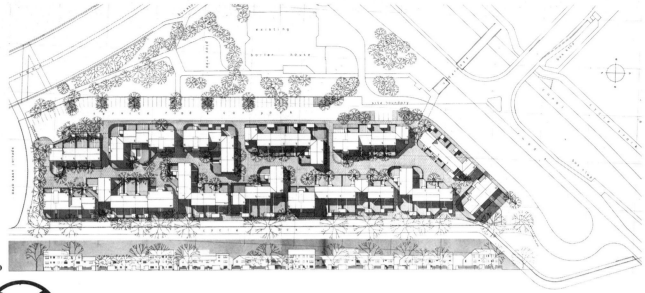

1.40

THAMESMEAD, KENT (AREA 3L)
Parking off perimeter road; no cars within housing area
(Architects: Greater London Council Department of
Architecture and Civic Design)
In the third example of this type not only cars but also
service vehicles are excluded from the housing area. This is
made possible by the narrow shape of the site which enables
fire-fighting vehicles to be operated from the perimeter and
ensures acceptably short carrying distances to refuse and
removal vehicles, **1.40**.
The advantages of type 2 road layouts are:
1 The housing area is made safe for pedestrians and
  children at play
2 The housing area is free from disturbance from the noise
  of vehicles
3 The housing environment is not detrimentally affected
  by oil and exhaust stains and unsightly views of vehicles
  being repaired, caravans and boats in storage, etc
4 The architect is able to arrange the layout without the
  constraints of a road network to accommodate private
  cars.
The disadvantages are:
1 Greater walking distance from car to dwelling
2 Possible difficulties in provisions for service vehicles
3 Possible difficulties with regard to supervision against
  vandalism of parked vehicles.

EAGLESTONE, MILTON KEYNES, BUCKS
*Culs-de-sac penetrate housing scheme to parking courts; no
access from culs-de-sac except at end* (Architect: Ralph
Erskine)
At Eaglestone the architect wished to create a layout which
had a quiet and informal village character. One of the
methods used to achieve this was to prevent private cars
from circulating in the vicinity of the dwellings. The cars
are able to penetrate the housing area by means of short
culs-de-sac which branch out from the perimeter road,
1.41. These culs-de-sac are for vehicle use only; they have
no footpaths and there is no possibility of gaining access to
the housing area from any point along their length. Since
the driver has no incentive to stop until he reaches the
garage court at the end of the road, cars are not parked on
the road and so the full width of 5m is always available for
vehicle access.
The garages are situated around the perimeter of the
rectangular vehicle courts. Openings in these structures
provide pedestrian access into the housing areas. Hard
standings for visitors' parking are provided in the centre of
the courts so that cars parked in the open are screened from
the neighbourhood of the dwellings by the garage buildings
and perimeter walls.
A separate pedestrian network is incorporated in the
scheme with pedestrian underpasses at the perimeter road.

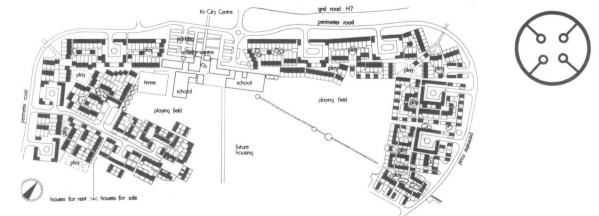

1.41

The advantages of the type 3 layout are:

1 The advantages of type 2 layout are maintained. These are maximum safety, freedom from noise, absence of the various types of pollution which accompany the car and the freedom of the architect to arrange the layout without the constraints of a road network

2 The use of access cul-de-sac reduces the walking distance from car to dwelling

3 The culs-de-sac provide service vehicle access to the vicinity of the dwellings

4 The cost of the culs-de-sac is reduced since pedestrians are normally excluded and the common, expensive provision of two footpaths, each 2m wide can be dispensed with.

The disadvantages are:

1 There remains the need to walk a short distance from the car to the dwelling

2 There is a possible difficulty with regard to supervision against vandalizing of parked cars

3 It will probably still be necessary to provide access for emergency vehicles beyond the ends of the culs-de-sac.

A typical view out from the housing area to a garage court is presented in **1.42**. The containment of the car means that children can move freely in the area of their homes without being in danger from moving vehicles. Play areas have been liberally provided among the houses, **1.43**.

1.42

1.43

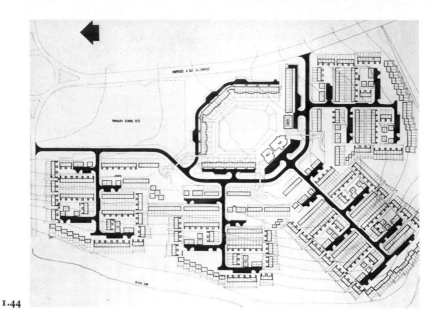

1.44

ANDOVER, HANTS

Access by means of a system of culs-de sac (Architects: Greater London Council, Department of Architecture and Civic Design)

As with other types of layout, many different interpretations are possible when using a system of culs-de-sac.

The one illustrated, using straight spine access roads with shorter culs-de-sac branching off at right angles, is probably the arrangement most commonly adopted in the recent past, **1.44**.

In this example the spine roads are 5.5m wide with short lengths widened to 7.5m to allow parallel parking at the side of the road.

The branch culs-de-sac are 4.88m wide and have an adjacent parking strip 5.5m wide to allow parking at right angles to the road.

1.45

HALTON BROW, RUNCORN

Access by means of a system of culs-de-sac (Architects: Runcorn Development Corporation, Architects' Department)

At Halton Brow in Runcorn New Town the designers set out to provide a system of access ways which were pedestrian orientated. The scheme was designed in 1966 and has attracted a great deal of interest since completion. The part of the development on the west side of the Busway (see **1.45**) is called The Calvers. This area is served by a local distributor 6m wide, which in turn serves four culs-de-sac of which the road width varies between 3m and 6m. These roads are designed as one lane, two-way roads with passing places.

The culs-de-sac serve small courts containing the parked cars, the houses being grouped round these courts. There is no frontage access onto the local distributor or the culs-de-sac.

As can be seen on the layout plan, the cul-de-sac access roads have been given an informal, organic shape. They have a meandering sett edging and no adjacent footpaths. As a result the road network ceases to be obtrusive and becomes an integral, enhancing element of the residential environment.

The sight lines and radii have been kept to a minimum appropriate to the design speed of 10mph.

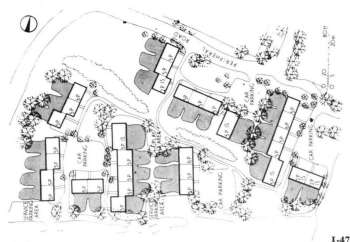

1.46

1.47

The large-scale plan, **1.46**, illustrates the northern section of The Calvers, as shown on the layout plan, **1.45**.

This plan indicates the relationship between roads, footpaths and dwellings.

The dimensions of the parking areas vary, but a typical size is about 17 × 15 m.

The serpentine alignment of a typical access road is clearly seen in **1.47**. This is the cul-de-sac which appears nearest the bottom edge of the layout plan. With

1.48 it illustrates the absence of paths along the road and the way in which the ground is contoured to rise steeply from the edge of the road. The narrowness of the carriageway discourages drivers from parking directly on it since this would block the single lane and incur the displeasure of neighbours. At the same time it is not possible to park with part of the vehicle off the road surface because of the steep side-slopes. These two factors, together with a third disincentive, effectively

discourage drivers from parking on the road. The third consideration is that the most convenient place to park is in one of the parking areas since this provides the most direct access to the dwellings. One of the reasons that the designer omitted the footpaths was in order to help make the dual-use nature of the surface clear to the motorist. However, as can be seen on the layout plan a separate network of footpaths has been provided for the convenience of pedestrians.

1.48

1.49

1.51

1.49 is a view of one of the car parking areas off the access road, while **1.50** shows a service vehicle parking area which has been provided by widening the access road locally.

Photograph **1.51** shows the unobtrusive edge detail in a parking area. In this instance the front entrance to the dwelling leads directly onto the parking surface. In other arrangements the dwellings face onto a footpath which leads to the parking area.

This edge detail is also used on the access road, **1.52**, where the designer's concern to ensure that the roadworks blend with their surroundings as much as possible can also be seen in his decision to have light coloured chippings rolled into the asphalt surfaces. The proximity of the road surface to structures, combined with sharp curvature of alignment and narrow width, inhibits excessive speed.

1.50

1.52

27

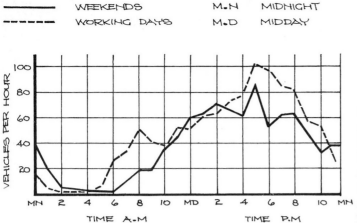

WEEKENDS   M.N   MIDNIGHT

------ WORKING DAYS   M.D   MIDDAY

1.53 *Average traffic flow generated by 208 dwellings at The Calvers, The Brow, Runcorn, Cheshire (Source : E. Jenkins, 'Highway Hierarchy—or please don't bring your car into the living room',* Institution of Highway Engineers Journal, *November 1975)*

Some designers have misgivings concerning the use of one lane, two-way roads, fearing that excessive delays and inconvenience will result. It is therefore of interest to note the result of some traffic counts which were carried out on this development. About 1040 vehicle movements were generated by 206 dwellings on an average working day. During the peak hour, between 5 pm and 6 pm, only about 100 vehicle movements occurred, **1.53**. That is approximately 25 vehicle movements per cul-de-sac, or less than one every two minutes during the peak hour. At the time of the traffic survey 50% of families owned cars. Clearly even if the car ownership rose to 100% this road system could easily cope without inconvenience to pedestrian or motorist.

The designers' objective of producing a small-scale, informal system in which the motorist would be induced to drive slowly and with caution has been realized. The design was a result of a sympathetic cooperation between architect, engineer and landscape architect, each member of the team having the same goal of creating a relaxed residential environment on a human scale.

Halton Brow has a site area of 10.7 hectares (26.5 acres) with 356 dwellings—a density of 126 persons per hectare (pph), (51 ppa).

The Chief Architect and Planning Officer of the scheme was R. L. E. Harrison, in succession to F. Lloyd Roche, with P. Riley as Principal Architect responsible for housing and R. N. E. Higson as Principal Architect responsible for landscape. The Chief Engineer was J. Mercer and the Principal Site Development Engineer was E. Jenkins.

LANGDON HILLS, BASILDON, ESSEX

Access by means of a system of culs-de-sac (Architects: Basildon Development Corporation, Architects' Department)

At Langdon Hills the culs-de-sac are aligned at right angles to the main access road, **1.54**. An irregular curved alignment has been chosen. Straight roads can encourage high speeds; by using a curved alignment speeds are restricted and a constantly changing view along the road results. The exact configuration of the curved roads was determined to a large extent by the need to avoid existing trees which the designers decided to keep as they were an important asset of the site.

In effect these culs-de-sac take the form of a landscaped close incorporating parking areas. Every effort was made by the architect to reduce the formality and 'hardness' of the road and parking spaces.

1.54

The road curvature in association with straight terraces produces useful areas for screened car parking, sitting-out areas and children's play, **1.55**. Provision for emergency vehicular access through the terrace has been made at critical points.

In this layout the domain of the pedestrian and that of the car are kept separate, **1.56**. Pedestrian routes are placed parallel to the terraces with a low wall between the footpath and the road and parking areas. In this way the resident on foot need not follow the serpentine route of the vehicles. An added advantage is that the young children's play areas which are located adjacent to the footpath can be reached from the dwellings without approaching a road, **1.57**.

55

56

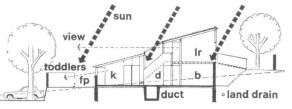

57

29

Langdon Hills has a site area of 13.37 hectares (33 acres) with 556 dwellings—a density of 202 pph.

There is 150% parking provision, with spaces for 834 cars. The Project Architect was Clive Plumb.

The advantages of type 4 layouts are:

1 Cars and service vehicles can penetrate to the vicinity of the dwellings.

2 Culs-de-sac have a better safety record to date than any other type of access road (of course this type of layout cannot be expected to be more safe than types 2 and 3).

3 There is no vehicular through traffic.

The disadvantages are:

1 Some danger to pedestrians and children at play exists

2 Residents are subject to the various types of pollution associated with cars

3 Possibility of residents on culs-de-sac experiencing a sense of isolation

4 Precautions must be taken to obviate the potential seriousness of a blockage to the route

5 Probably greater road costs than with types 2 and 3.

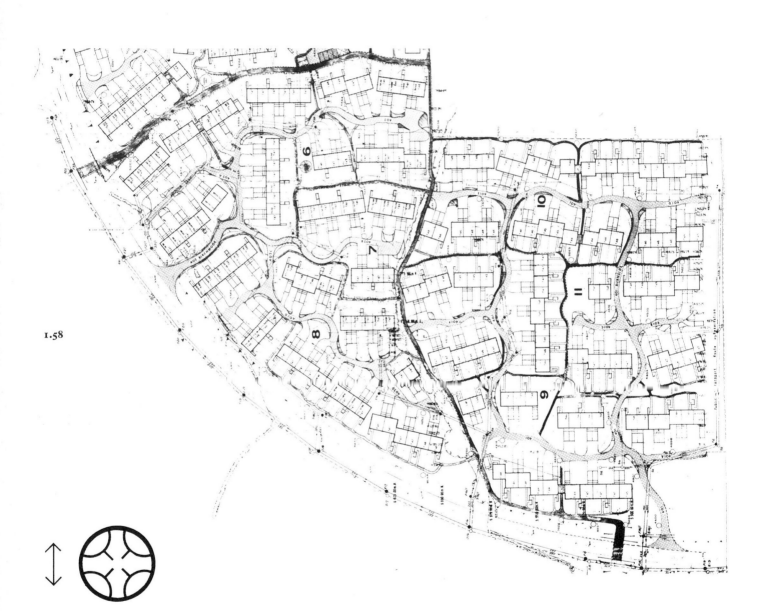

1.58

BROOKSIDE 4, TELFORD, SALOP

Loop roads in combination with culs-de-sac (Architects: Telford Development Corporation, Architects' Department)

The right-hand side of the layout plan of part of Brookside 4, **1.58**, illustrates the case of a loop road being used in association with branching culs-de-sac. The serpentine access roads in this development are 2.75m wide. The routes operate as two-way single lane roads, the forecourts of the paired garages acting as passing places and in some locations taking the place of hammerheads and turning circles. The device of using the garage forecourts in this way has the advantage of ensuring that the passing places are not used for parking.

The curved alignment and narrow width of road induce low speeds and caution.

1.59

HALTON LODGE, RUNCORN, CHESHIRE
Loop roads in combination with culs-de-sac (Architects: Runcorn Development Corporation, Architects' Department)

Halton Lodge is another example of a layout using a loop road and cul-de-sac combination, **1.59**.

In this case a variable road width has been used. The basic width of 5 m constitutes a two-way two lane road. The road is widened at strategic points to between 6.7 m and 7.3 m.

The advantages of type 5 layouts are:

1 Turning heads are not necessary unless loop roads are very long
2 It is possible in certain circumstances to use one-way, one lane roads
3 Cars and service vehicles can penetrate to the vicinity of the dwellings
4 A blockage of the route is not as serious as on a cul-de-sac.

The disadvantages are:

1 If one-way loop roads are used some drivers may be inclined to traverse parts of the road in the wrong direction to avoid a long detour
2 Unless the roads are designed with care drivers may be more inclined to speed on loop roads than on culs-de-sac
3 Care must be taken to ensure that the loop road is not an attractive detour to through traffic.

Note: Disadvantages 1 to 3 of type 4 networks also apply to type 5 (see p.30).

1.60

CHRISTCHURCH SITE, VICTORIA PARK, HACKNEY, LONDON
A number of enclosed areas connected in series (Architects: John Spence and Partners; Developers: Crown Estate Commissioners)

In this intelligent and sophisticated development the designer has used a variety of methods to provide vehicle access. A cul-de-sac terminates in Vicars Close, Pennethorne Close leads to a mews court and St. Agnes Close ends in a dual use area. Pedestrian precincts are located between these areas. The result is a number of spaces linked in series providing a neighbourhood of great interest and variety, **1.60**. Other spaces linked in parallel add to the diversity and choice of route for the pedestrian.

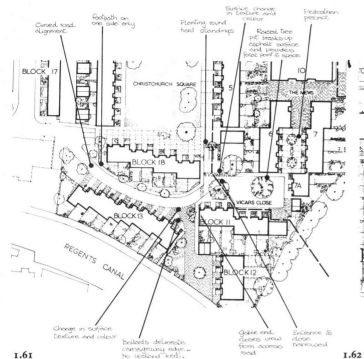

Curved road alignment.

Footpath on one side only

Planting round hard standings

Surface change in texture and colour

Pedestrian precinct

Raised tree pit breaks up asphalt surface and provides focal point to space.

BLOCK 17

CHRISTCHURCH SQUARE

THE MEWS

BLOCK 18

VICARS CLOSE

7A

BLOCK 13

BLOCK 11

REGENTS CANAL

BLOCK 12

Change in surface texture and colour

Bollards delineate carriageway edge.— No upstand kerb.

Gable end closes view from access road

Entrance to close narrowed

1.61

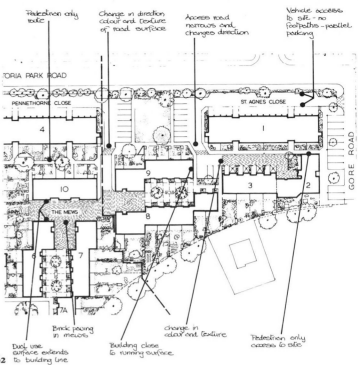

Pedestrian only route

Change in direction colour and texture of road surface

Access road narrows and changes direction

Vehicle access to site – no footpaths – parallel parking

VICTORIA PARK ROAD

PENNETHORNE CLOSE

ST. AGNES CLOSE

GORE ROAD

THE MEWS

Brick paving in mews

Dual use surface extends to building line

Building close to running surface

change in colour and texture

Pedestrian only access to site

1.62

The access road to Vicars Close from Northam Street is gently curved over most of its length and is then sharply curved before entering the close. As a result the view along the access road is terminated by the gable wall of block 11 until the sharp bend is reached and then the close becomes visible. Thus the road curvature has been used to introduce the element of surprise, to provide privacy to the close and to induce caution in the driver. A footpath has been provided on one side of the road only, **1.61**.

The access to the mews at the end of Pennethorne Close and at the end of St. Agnes Close have a serpentine alignment, **1.62**. This factor together with road

narrowing and frequent changes in surface texture ensures that cars are driven at walking pace in the mews courts.

1.63 depicts Vicars Close at the end of the curved cul-de-sac access road. The ubiquitous hammerhead has been avoided and a unique place has been created. The proximity of the front doors of the dwellings to the road surface heightens the urban character of the space.

This relaxed, informal close leads into a pedestrian-only precinct through the gap in the low wall to the left of the photograph.

This precinct provides a safe area for children's play and a quiet place away from the noise of traffic. It leads, in turn, into a mews court through the gateway in the high

wall seen in the middle distance of **1.64**.

The brick paved surface of the mews, **1.65**, is in contrast to the normal asphalt surface of roads for traffic. This has the effect of inducing a feeling of caution in the driver as does the sense of enclosure which has been created in the space.

The entrances to the dwellings have been recessed in order to provide a small outdoor space which acts as a transition between the inside of the building and the outside area which can be used by cars as well as pedestrians, **1.65**. The Mews leads to another pedestrian precinct opening into St. Agnes Close and the east entrance to the site.

1.63

1.64

.65

The advantages of type 6 layouts are:
1 By the use of a variety of means of access it is possible for the designer to create an environment of great variety and interest.

The disadvantages are:
1 The designer must take great care to integrate all the elements of the design into an organic whole.

Note: Advantages 1 to 3 and disadvantages 1 to 5 of type 4 layouts may apply to type 6 layout.

HOUNSLOW HEATH, HOUNSLOW, MIDDLESEX

A grid of interconnected areas (Architects: Greater London Council, Department of Architecture and Civic Design)

This design for a large housing development on Hounslow Heath is of considerable interest from the point of view of road access. The scheme was unfortunately not built since it was decided that the noise from aircraft using the nearby Heathrow airport reached unacceptably high levels. Nevertheless the design is presented here since it is an excellent example of the integration of the road access system into the total concept of the housing area.

Before considering the road system it is appropriate to examine the aims of the designers, because the type of road network chosen was an integral and indispensable part of the design solution which was developed.

The statement by the architects which is reproduced below summarizes these aims.

Planning and design

Initial studies indicated that in planning a large development on Hounslow Heath consideration must be given to the following problems:

A large flat expanse with few natural features, the Heath is isolated from the remainder of Hounslow. The new development would be mainly low rise housing with a relatively small amount of public building. Unless monotony was avoided and a good transportation system developed, residents would have difficulty in identifying their own part of the scheme and in relating to both the remainder of the Heath and the community beyond.

The proposed plan included the following features:

1 Landscaping of the entire Heath would establish it as a local amenity as well as provide a setting for the new housing.

2 New bus routes into the development would offer easy access to the centre of Hounslow for major shopping facilities and transport to central London on the Piccadilly Underground. In addition the buses and roads would enable the people of Hounslow living outside the Heath to share the new community's facilities.

3 Within the whole development of over 2,000 homes there would be a series of smaller groupings to which a house could be seen to belong and with which the individual families could identify.

The development was divided into four Neighbourhoods grouped around a landscaped open space and lake. The main public buildings would be arranged together with some higher density housing on one side of this central area. The neighbourhoods of some several hundred dwellings would be separated by fingers of open space connecting the central green with the surrounding Heath.

—Each neighbourhood was to consist of several loop access streets off a central spine road. About 100 dwellings would share a single loop street.

—Along each loop street clusters of about 20 to 30 dwellings would be arranged around squares. Here cars would be parked and pedestrian and vehicle routes meet.

—Each dwelling would belong to a small number of houses built around a pedestrian space; up to four or five of these groups would define a square.

4 Closely related to this housing plan would be a system of open spaces beginning with the private garden, patio or roof terrace provided for each dwelling and enlarging first to the local green within each loop street, then to the central green and lake and finally to the whole Heath itself.

Roads and footpaths

Traffic within the scheme was to be controlled by a hierarchy of roads each with a distinct function and a limited speed.

1 Vehicle entry to the scheme would have been from either Staines Road or Wellington Road. In each case the road would first cross the open Heath on its approach to the central area.

2 Each neighbourhood would be served by spine roads leading off the main site access roads. Along these roads a series of central islands forming 'chicanes' would reduce vehicle speeds to about 25mph.

3 Access to dwellings would be via a one way loop street off the spine road. These narrow loop streets were designed with sharp bends and surfaced with concrete slabs to ensure traffic speeds of less than 15mph.

The footpath and road systems would coincide in the access squares which were designed to cater for a variety of activities. Otherwise the footpaths would pass diagonally through the housing areas to the central green and community facilities and would link the open spaces.

Housing

The form of the housing was designed to satisfy a number of criteria:

1 The height of the buildings should not exceed three storeys over most of the scheme and four or five storeys along the boundaries to the Heath and the central green where a higher density was required.

2 As many dwellings as possible should have gardens, patios or terraces.

3 The building groups should enclose the squares and define the pedestrian spaces.

4 A variety of form and treatment would be essential to avoid monotony and to enable people to distinguish between different parts of the development.

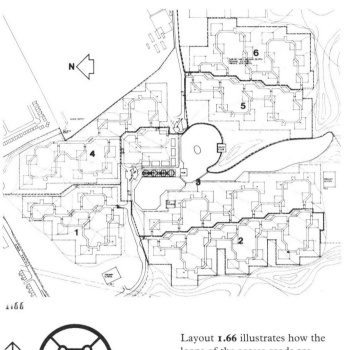

1.66

Layout **1.66** illustrates how the loops of the access roads are interlinked in each neighbourhood to form a grid network.

The access routes which connect the clusters of dwellings and which join into the access collector roads (those with the chicanes) were planned as one-way streets. It was intended to provide raised footpaths on at least one side of these streets. The squares on the other hand are designed as completely dual-use areas. In these areas the alignment of the route for through traffic is marked by the use of a different colour or shade of surfacing material and by the use of a different coloured flush edging strip, **1.67** and **1.68**. The relationship of the dwelling layout to the road network can be seen in the plan of the north east neighbourhood, **1.67**.

As can be seen, the devices to ensure that the design speed would not be exceeded include the incorporation of continuous changes of direction and the close proximity of the running surface to the building line at many points.

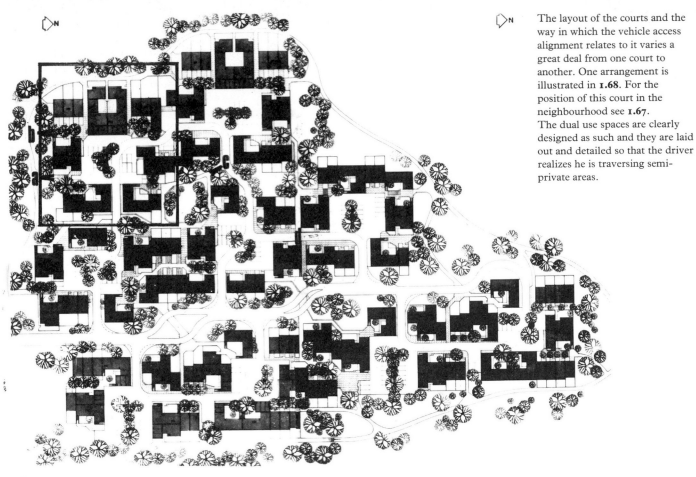

The layout of the courts and the way in which the vehicle access alignment relates to it varies a great deal from one court to another. One arrangement is illustrated in **1.68**. For the position of this court in the neighbourhood see **1.67**.
The dual use spaces are clearly designed as such and they are laid out and detailed so that the driver realizes he is traversing semi-private areas.

1.67

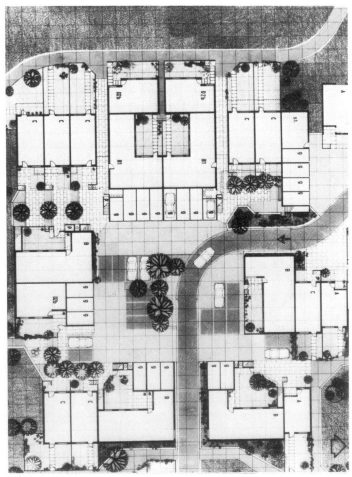

1.68

The advantages of type 7 layouts are:
1 The flexibility of the arrangement creates new possibilities for the layout designer
2 The less formal vehicle access system can be exploited to reduce the impact of the roads on the environment
3 Dual use areas can create the opportunity for cost economies.
The disadvantages are:
1 Lack of clarity in the form of the layout may result if particular care is not taken over this aspect of the design.
Note: Advantage 1 and disadvantages 1 and 2 of type 4 layouts also apply to type 7 layout.

35

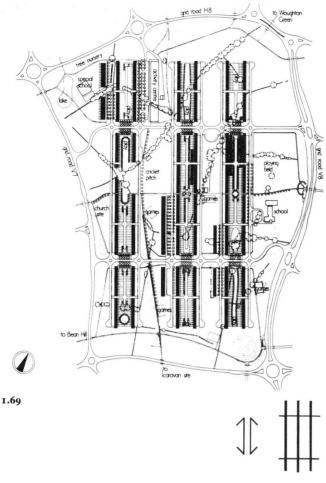

1.69

The width of the access roads parallel to the terraces is 5.5 m with 1.8 m wide footpaths where the dwelling faces on to the road. Those at right angles to the lines of the terraces are 6 m wide with two 1.8 m footpaths set back from the road.

Although Netherfield has been used to illustrate a grid form of layout, this type or road pattern can be used in conjunction with a much tighter arrangement of buildings. This is especially the case since the Department of the Environment published its recommendations for the layout of residential access roads contained in Design Bulletin 32.

The advantages of type 8 layouts are:

1 Dwellings can be reached with maximum speed
2 The logic of the layout is clearly apparent.

The disadvantages are:

1 The layout can be monotonous
2 The layout can become out of scale for the pedestrian
3 There is a danger of cars being driven at excessive speeds
4 Cars are likely to be parked on the access roads. These are a major cause of accidents involving children in housing areas.

NETHERFIELD, MILTON KEYNES, BUCKS
A grid of interconnecting access roads (Architects: Milton Keynes Development Corporation, Architects' Department)

In this layout the designer has adopted a traditional rectangular grid network with straight terraces parallel to the road, **1.69**. A paramount consideration was that each dwelling should have easy access by car. The scale and appearance of the housing area proclaim the fact that the needs of the car have been given a high priority.

As can be seen from the layout drawing and from **1.70** the dwellings have been constructed on both sides of the access roads only along comparatively short lengths. This relieves what would otherwise be very monotonous routes along the line of the roads and provides the occupants of most dwellings with a view over the road to a landscaped open space.

In some areas parking space is provided adjacent to the road by an extended widening of the surface area. In other parts of the development individual hard standings are provided in front of each dwelling.

1.70

General observations

A comparison of the types of road access layouts presented brings out some points worthy of mention.

1 The type of access road arrangement has a fundamental effect on the layout and environment of the housing area which is developed by the designer.

2 The access road layout should satisfy a number of criteria. The degree of importance allocated to each of these criteria is a critical factor in the selection of a particular type of layout. The most relevant factors in terms of the needs of the residents are listed below.

   (a) The need to have an environment of integrated built forms to a human scale. This need calls for road design which has a unifying effect on the layout and not a devisive one.

   (b) The need of residents to be able to circulate on foot within the housing area, in pleasing surroundings, without being in danger from moving vehicles and without having to walk excessive distances.

   (c) The need of residents to be able to drive to within a convenient distance of the dwelling.

   (d) The need of service vehicle operators to be able to drive within a convenient distance of the dwellings.

The order of priority and the precise interpretation given to these needs will have a significant effect on the type of road layout used.

In the case of (a), if the unifying effect of the road is allocated a major priority then a designer might select a series of dual use courts. If this is given a low priority, a rectangular grid might be used.

If (b) is considered of major importance then the car may be excluded from the housing area.

If (c) is to be top priority then the grid network may be selected or some other arrangement of dwellings lining access roads with adjacent hard standings.

In the case of (d) the statutory undertakers will probably have definite requirements which must be complied with. Nevertheless, some leeway is often possible and detailed discussions between the designer and the statutory undertakers can often lead to a solution which most satisfies the needs of the residents. Of course, the service vehicles do not necessarily have to use the same routes as private cars.

The interpretation of the needs listed is subject to the requirements dictated by the physical situation of each particular housing area and the wishes of its inhabitants, if these can be determined. No list of priorities with regard to people's preferences in these respects can be compiled. A designer must develop the best solution for each individual case.

There are, of course, many other factors which the designer must consider in the selection of a road layout type. These include types of parking provision, degrees of privacy, grouping of dwellings, types and uses of spaces.

2
ELEMENTS OF ALIGNMENT

We are all familiar with those arrangements of roads and buildings which give special delight: the city squares and crescents; the mews, courtyards and alleys of towns; the village green and lanes. In our preoccupation with the efficient circulation of vehicles many of these indigenous elements of townscape alignment, which so greatly contribute to the variety and individuality of our urban areas, tend to be overlooked. A simple résumé of traditional elements is presented here in order to bring to the attention of the designer the wide range of possibilities.

Straight and curved roads

From the environmental standpoint the important prospects of a road are those obtained by a person viewing it from ground level, whether he is on foot or in a vehicle. What matters is the way the place or route is experienced by the user. The view from the drawing board or the aeroplane is only significant to the extent that it has a bearing on this ground eye view.

Having made this reservation it is interesting to compare the two aerial photographs, **2.1** and **2.2**. The linear elements are emphasized in **2.3** and **2.4**. The impression made by these two patterns is quite different and reflects some of the differences in character of straight and curved roads when seen from ground level.

In **2.1** the layout looks formal, orderly and strictly planned, whereas in **2.2** it appears relaxed, gentle and organic. This is not to say that straight and curved roads will of necessity result in environments with these attributes, since many factors go into making the total character of a place: the deprivation associated with an industrial revolution terrace and the affluence often associated with a low density surburban development are not, of course, the result of the road alignment. Nor is it to say that one type of road is always preferable to another. Nevertheless generally speaking curved lines appear more relaxed and organic and straight ones appear to have more tension.

STRAIGHT ALIGNMENTS

There is a long tradition for the use of straight roads in urban and residential areas. They have been employed in the creation of many different types of environment from the slums of the industrial revolution to expensive modern suburban developments. They have both advantages and disadvantages.

As is the case with all parts of the townscape environment the roads should be designed to a human scale. This means that unless it is being designed exclusively for vehicles a straight length should not be so long as to intimidate the pedestrian. He should feel that he can walk to the end without becoming tired or bored. As well as being dispiriting for the pedestrian, long straights encourage driving at high speeds.

The speed on curved roads can be controlled by the radius since the centrifugal force and the effort of steering round the bend can warn the driver if he is exceeding a safe limit. Thus the design speed is experienced directly by the driver. On a straight road with a perfectly smooth surface and a flawless vehicle the design speed would be infinite. The only physical inhibiting factors on a long straight, assuming the absence of other vehicles, are the irregularities in the surface and problems with steering due to

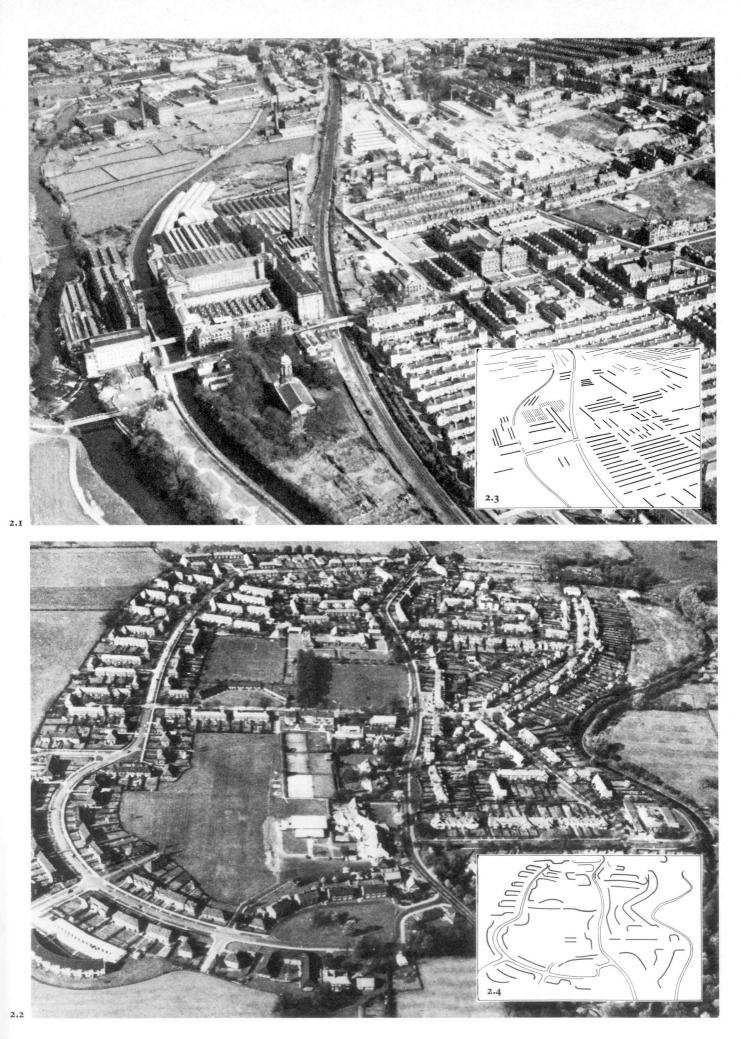

2.1

2.2

2.3

2.4

imperfections in the machine. These are much more difficult factors for the driver to judge.

There are a number of ways of discouraging high speeds by variations in the road surface. (These are discussed in Part IV, Materials, Surface and Trim.) In the example illustrated in **2.5** small roundabouts have been used at the point of intersection of the longitudinal and transverse elements of the grid.

An advantage of the use of straight lengths is that this can make it easier for people to orientate themselves in the area. When properly used they can impart a pleasing and easily perceived coherence to an urban area.

As well as being safe and to a human scale the road should be visually interesting. It should be a pleasant place to be in. Apart from the use of short rather than long lengths of straight there are a number of design considerations which, if given due weight, will greatly enhance the environmental quality of a straight road. These design considerations are examined below.

2.5

CLOSURE OF A STRAIGHT ALIGNMENT

The parallel edges of a straight road, if continued for a long distance, appear to meet in infinity. This effect is emphasized when the sides are contained by buildings or vegetation and a highly dynamic space results, **2.7**. If the line is blocked by a building, vegetation or landform the dynamic quality is arrested and a more relaxed, static space is achieved, **2.6** and **2.8**.

It is important to close off the end of a straight road in order to contain the space delineated by the surface and its sides. This helps to instill a sense of place into the linear space and to arouse a satisfying sense of completeness in the observer. This closing of the view is as important in the suburban road, **2.9**, as in the urban one. A condition in which it may not be desirable to enclose the end of the road is when the alignment leads towards a notable view which then may be exploited by the designer.

The most common method of attaining closure for a straight road is by means of the T-junction, **2.10**. This solution is doubly satisfactory since it works well in terms of traffic flow in the townscape alignment in addition to being a pleasing townscape form.

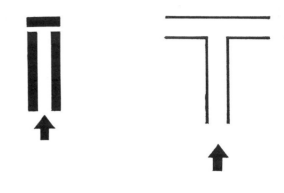

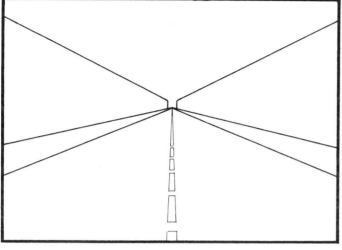

2.7

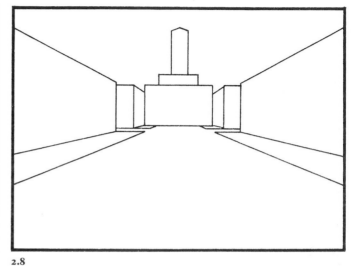

2.8

2.9

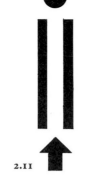

2.11

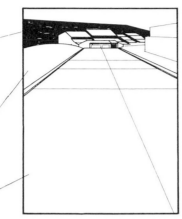

VEGETATION PROVIDES A
DARK BACKDROP TO THE
PRIMARY VIEW

ALIGNMENT OF THE ROAD
IS EMPHASISED BY THE
MOUNDING AND BY THE
BRICK WALL ON THE
OTHER SIDE
DISTRACTING SECONDARY
VIEWS ARE BLOCKED

ALIGNMENT OF ROAD
FOCUSES ATTENTION ON
MOST IMPORTANT VIEW
PARALLELISM IS AN
ASSET

2.12a

b

THE TERMINAL VIEW

The interest of the road can be greatly enhanced for the pedestrian or the driver if it has a focal point at the end which is being approached, **2.11**. In this way the parallelism which can be a main cause of monotony is turned into a virtue by focusing attention on the important view. Such a device is particularly valuable if the road otherwise lacks interest. In some cases it may even be desirable to screen or mask the side view in order to focus attention on the important terminal one. In **2.12a** and **b** all the features of the environment assist in directing attention to the far end.

It may also be justifiable to make the side view repetitive if the terminal view is strong enough to sustain interest.

In **2.12b** the terminal view is a building with an intriguing shape which could be only partly discerned from the far end of the road. Such a prospect arouses the viewer's interest and draws him forward to investigate further.

A different type of termination is the massive edifice which by its scale and drama, dominates the road from any position along its length, **2.13**. Such an arrangement puts an unequivocal full stop to the line. The destination is clear and its presence already pervades the place from far off.

At the other extreme the focal point can be quite a small sculpture or monument in which case some additional factor is required to complete the enclosure, **2.14**. In this example the trees of a small urban park fulfil this function.

2.13

2.14

The vertical alignment

The straight road, unlike the curved one, tends to reveal all its attractions immediately to view. Although the aspects of the containing sides alter as the route is traversed, little is held back. This state of affairs is modified if the alignment is on a vertical curve. The view towards the end of the road is more or less obscured depending on the degree of curvature. This holding back of part of the view can increase interest and lead a person on to explore further.

The road in **2.15** looks pleasing and interesting in spite of being contained on the sides by plain walls. Its visual success is contributed to by the vertical curvature, but also by the scale, the planting, the materials and the buildings glimpsed on the high ground to the right of the picture.

2.15

Containment of the sides

The road surface should work with the other elements of the environment (buildings, planting, landform) to structure space. In this way the surface becomes a positive element in creating spatial order. The forms at the sides of the road create the volumes implied by the plane of the road, **2.16**.

The nature of the containing forms at the sides of the space are of major importance in determining the character of the road. This aspect of road design is relevant to all roads, straight or otherwise, and is discussed in Chapter 3, Form, below.

Here we will refer to only one aspect of the form taken by the containing sides. This is an aspect which is especially relevant to straight roads and concerns the perceived 'speed' of the road.

2.16

2.17

The parallel lines bounding a straight road surface appear to converge rapidly to the horizon. If the lines of the containing built forms are predominantly horizontal they too will appear to converge towards the horizon, **2.17**. This results in a space which seems to rush the viewer to the opposite end. It appears

restless and aggressive.
The vertical lines of the windows and the columns in **2.18** tend to counterbalance the receding horizontals, yet the repetition of the identical portico diminishing in apparent size into the distance reinforces the horizontal lines so that the 'slowing down' process is only partially effective.

In **2.19** the strong vertical lines of the railings and of the windows greatly contribute to the complete counterbalancing of the horizontals, resulting in a space which is significantly more static and restful. The verticality of the windows on the side elevations is emphasized by being seen in perspective and by the white

paint in the window recesses. There are a number of other factors at work slowing down the space. The intermittent balconies at different levels and all the other irregularities in the elevations play their part.

2.18

2.19

The absence of enclosing elements in a built up area traversed by a straight road can result in a dispiriting prospect, **2.20**. This view will be improved when the trees have matured. One alternative to the complete containment of the road is to make the road space visually a part of an adjoining space, **2.21**. Here the texture and colour of the surface and the sympathetic edge detailing enable the road surface to lie comfortably beside the green. The space is contained by the vegetation on the left hand side and the road is experienced as part of the total volume.

Another example of the road being part of a larger space, **2.22**. This is a route for service vehicles only, in a highly urban setting. Although it is partly contained in a sub-space by the flanking brick walls the proximity and size of the surrounding buildings is such that the road is embraced by the larger volume.

The intermittent nature of the containing elements on the right-hand side road edge and the use of a post and rail fence over part of the length help to reduce the parallelism.

2.20

Although monotony and a tendency to encourage high speeds are dangers in the use of long straight elements of road this is not to say that their incorporation in new designs is always to be avoided. Many highly successful examples are to be found, especially in our cities.

As always the best results are obtained when the road length is considered as a unity. It should be orchestrated so that sequences of views and visual events unfold as the road is traversed. This can be done in a great variety of ways by various combinations of good townscape elements.

2.21

2.22

Chester Row and Chester Square, Westminster

One example of the selection of various ingredients and their combination into an immensely satisfying unity is the long straight length of road consisting of Chester Row followed by one side of Chester Square in Westminster, London, **2.25**.

A feature of this road which contributes to its distinctive character is the climax being near the middle of the length rather than at one end. This climax is a church steeple, a powerful vertical element which acts as a visual anchor to the horizontal road length.

2.23 represents diagrammatically the dominant attributes of the design, which is examined in more detail by means of photographs **2.26** to **2.33**; the location of these is shown in **2.24**.

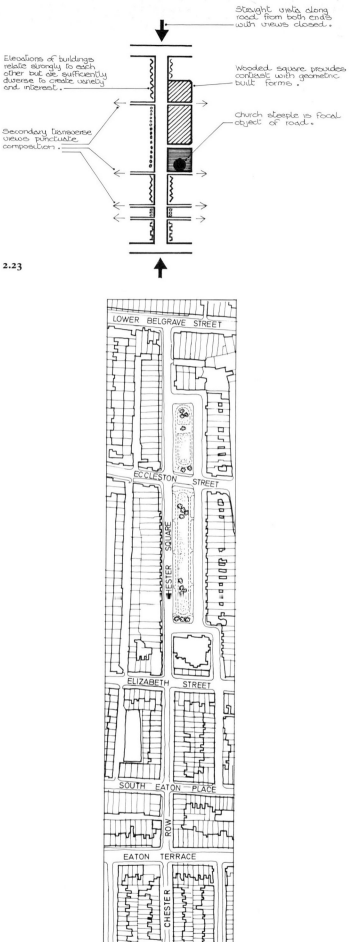

Straight vista along road from both ends with views closed.

Elevations of buildings relate strongly to each other but are sufficiently diverse to create variety and interest.

Wooded square provides contrast with geometric built forms.

Church steeple is focal object of road.

Secondary transverse views punctuate composition.

2.23

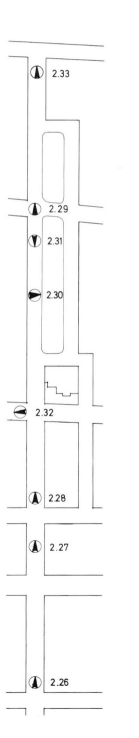

2.24

LOWER BELGRAVE STREET

ECCLESTON STREET

CHESTER SQUARE

ELIZABETH STREET

SOUTH EATON PLACE

CHESTER ROW

EATON TERRACE

CHESTER

BOURNE STREET

2.25

48

At the southern end of the road, **2.26**, the dwellings are built to a small scale which is accentuated by the small shop fronts, low sky line, fluctuations in building height over short lengths and frequent variations in detailing. Even at this considerable distance the church spire can be seen over the eastern elevation of the street. As progress is made along the road the scale of the buildings gradually increases, and at the same time the elevation becomes more formal, the formality being enhanced by the substantial symmetry of the two sides, **2.27**. This symmetry, although providing a fine approach to the climax of the spire, would become tedious if maintained from one end of the road to the other and indeed the arrangement becomes asymetrical some distance before the church is reached.

The balconies, wrought iron railings, tiled thresholds to front doors and many other carefully considered and executed details afford decorative elements which greatly add to the visual interest of the scene.

2.26

2.27

Further on the church becomes a more prominent feature as the façade on the right-hand side becomes more varied, **2.28**. Small shops in this mid-way area cater for the minor purchases of the local residents when they do not need to avail themselves of the more comprehensive facilities at the end of the road.

The canopy of mature trees jutting out in the middle distance invites the viewer forward to explore this planted area which provides a welcome contrast with the hard surfaces elsewhere. Past the church the uniform height of the buildings on the left-hand side, **2.29**, together with that of the structure straight ahead closing the view, defines the ceiling of the road space. The enclosing effect of the façade at the north end of the road and at right angles to the road line begins to make a more powerful impact. This enclosure of the road line is maintained on the right-hand side by the trees, which introduce an informal element into the composition although the planting is strictly contained within the rectangular area defined by the grid arrangement of roads.

2.28

2.29

50

The planted area of Chester Square is a quiet, secluded retreat; a semi-private environment in strong contrast to the public domain of the road, **2.30**. Although the boundary line of this area is a formal part of the overall plan pattern, and the boundary hedging is carefully formed by clipping, the other aspects of the planted area are informal, thus contributing to the relaxed atmosphere of the place. An important contribution to the character of the road space is made by the particular relationships between road, building line, and access to the front of the dwellings. In spite of the proximity of the building line to the pavement a semi-private zone has been created which enables a gradual transition from inside to outside. The front doors are slightly recessed and are protected by two columns and an overhanging balcony, **2.31**; on emerging from a dwelling it is necessary to take a few paces and descend three steps before the pavement is reached.

These measures on the part of the designer increase the sense of privacy in such a highly urban and densely populated area. As well as distancing the user of the dwelling from the casual passer-by, the steps elevate the ground floor above the general level of the street enabling those looking out to see over the railings and cars and making it more difficult for those outside to see in.

The existence of the basement, with its light well between the pavement and the building line, constitutes a further barrier between inside and outside and provides a second, informal entrance to the dwelling from the street.

A glance back down the road, **2.31**, shows the focus of the design, the church spire, partially concealed by the tree canopies as it was partially concealed on the opposite approach by the building elevations. This is an invitation to the viewer to move further down the road and explore behind the obstruction.

The long view to the end of the road in this direction focuses on and is finally blocked by the distant trees.

The rectangular grid layout means that each length of road is intersected transversely by other lengths, **2.32**. These sudden changes of view provide a dramatic punctuation to the journey. Of course, if the transverse views are identical to that straight ahead this will produce a tedious and disorientating effect, but if they are not the same they can be designed as complementary to the constant view and will enrich the visual experience of the observer. Finally in **2.33** the main line is unequivocally blocked by the static proportions and disposition of the Georgian block.

2.30

2.31

2.32

2.33

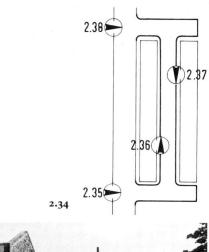

2.34

Lofting Road, Islington

To be successful, straight roads need not be on the grand scale of Chester Row. An example on a much smaller scale, but equally accomplished, is that of an infill development at Lofting Road in the London Borough of Islington. This road provides vehicle access between two rows of dwellings, **2.34**.

It is designed on the basis of a two-way loop. The width of the running surface is only 2.5 m but there is ample room for passing opposite the integral garages.

One of the two entrances, **2.35**, showing the cantilever-arm gate which enables the access to be closed to vehicles when desired. The entrance to the main length of the route parallel to the building line is inconspicuous. This promotes a slow vehicular approach, as does the slight ramp up and the rough textured strip of surfacing crossed on entry.

The straightness of the main length of road is balanced by the sculptural treatment of the structures lining the route, **2.36** and **2.37**. It is clear from the photographs that the central space has been designed as a total concept and not as a road which would subsequently have buildings lining the sides or as rows of buildings which would require some form of road to be added as a final design consideration.

By the use of setbacks, projections, recesses and so on, the volumes and mass of the buildings become apparent, creating variety and interest and at the same time breaking up the parallelism which would have resulted if a straight terrace with a continuous, flush elevation had been used. The skyline too has been varied but not sufficiently to destroy the unity of the whole. The three plan elements of the road, one parallel to the longitudinal axis of the site and two at right angles to it, are all contained in clearly defined volumes. The views are closed and the building heights on all sides are sufficiently uniform to clearly indicate the ceiling of the space.

2.35

2.36

37

8

The central brick gutter and the transverse strips of brick surfacing soften the impact of the asphalt surface, **2.37**. A further softening effect on the predominantly hard-landscaped, urban character of the development is provided by the small but strategically placed areas of planting. These have not yet had time to become established but will eventually make a significant contribution. It

is apparent that as much careful attention has been given to the floorscape as to the vertical surfaces, and the high quality of detailing greatly contributes to the success of the spaces.

The unity of the surface is greatly enhanced by the use of a minimal kerb upstand of only 25 mm, rather than the more usual 100 mm, and by employing a dark stone kerb. Differences in texture, colour, module of unit and the

use of line and area to subdivide the total surface have all been combined towards the end of designing a pleasant and unique place. The bollards introduce a vertical element in the floorscape and at the same time fulfil the practical functions of indicating the positions of the parking bays, ensuring that cars do not cut diagonally across the pavement when entering or leaving a parking space and preventing

vehicles travelling the length of the route from mounting the kerb, except when this is necessary for passing.

The other access area, **2.38**, shows the inconspicuous junction with the main access road.

The device of placing the front door at right angles to the road rather than facing it provides a more sheltered position when the building line is close to the road, and enhances the sense of privacy.

53

2.39

CURVED ALIGNMENTS

It has been mentioned that, because of the centrifugal effect, the use of a curved alignment enables the designer to exert control over the speed of the vehicles. It is also the case that the driver has more control over the vehicle when following a curved path than he does on a straight one.

Vehicles in motion, then, benefit from the mathematical properties of curved lines. It is not so apparent why people on foot should prefer to follow a curved path rather than a straight one, but there is evidence that this is the case. Paul Jacques Grillo, in *Form, Function and Design*,[3] shows two photographs of a flat grassed area on the campus of the College of Notre Dame, California. One picture depicts the summertime view of a perfectly straight paved pathway running from one corner of the flat area, straight across it to an opposite corner. The other photograph shows the same scene in wintertime when the whole area is covered in a few inches of snow. The path taken naturally by the students is quite visible from the tracks and it can be seen to be a very pronounced reverse curve. In that case at least the straight line of the T-square and the theodolite was irrelevant to the natural inclinations of the walkers.

2.40

There is no doubt that the strict geometry of **2.39**, although not without its visual interest, is much less enticing than the relaxed line of **2.40**. As this photograph illustrates, the most gentle curve when seen in perspective is quite distinct. And even if time saving were to be a criteria the advantage of making the line in **2.40** straight would be completely negligible.

One of the attractions of the curved path for the person on foot is probably that wending along a route makes it easy for him to survey much of his surroundings, whereas a straight path focuses his attention on one point in the distance.

54

Closure

On a straight road in the absence of a vertical curve it is possible to see from one end to the other. On a horizontal curve which is sufficiently pronounced the view is completely closed, **2.41**.

If the curve is approached by a straight, a sense of anticipation is experienced, **2.42**, as the curve draws near and is entered, and a new prospect begins to unfold.

When the horizontal curvature is sufficiently gradual the main prospect may be already within sight from the start of the curve. Then the curve will slowly alter the orientation of the viewer with the accompanying promise of the vista expanding or of a different view being eventually revealed when the end of the curve is reached, **2.43**.

The vertical lines of the window openings and the progressive stepping down the slope of adjacent sections of the terrace provide a vertical emphasis which balances the onward sweep of the curve.

Changing views

As the pedestrian or driver moves along the curved alignment his view ahead is continually changing. We have pointed out the desirability of designing a straight length of road as a complete unit, with each part relating to the other parts and the whole constituting a unique composition; the same consideration applies to other types of alignment, and the curved line is particularly amenable to the application of this concept. The road can be organised as a series of gradually changing, inter-related views, **2.44**.

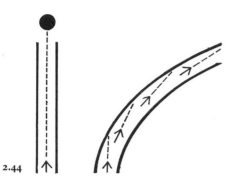

2.44

The pronounced curvature of some formal crescents necessitates the use of specially adapted floor plans, **2.45**. But since on a circular curve the same plans can be repeated round the crescent, the associated constructional problems may not be as difficult or costly as they at first appear.

Less formal arrangements often use the juxtaposition of a series of rectangular blocks to negotiate the curve, **2.46**. The junctions of such blocks or the outside spaces created by their proximity have been the source of countless incidents of attractive townscape.

2.45

2.46

Protection against weather
Continuously curved roads or roads with bends at either end have the considerable advantage of providing an additional protection against the elements by blocking off the wind.

In **2.47** the bend has been turned simply by a rectangular building set at a sharp angle on the left side of the road, with a gap between angled blocks making space for a secondary approach road on the right-hand side.

2.47

SERPENTINE ALIGNMENTS

Serpentine alignments are composed of a series of reverse curves. We have suggested that people may have a natural preference for moving along a curved path. The need to use reverse curves to cover more than the shortest routes would be a consequence of such a preference, since over a long distance a single curve would have to be so flat as to be virtually a straight line otherwise an unacceptably long increase in the journey time would result.

A series of reverse curves is a line which affords maximum surveyance of the surroundings, as well as being a reasonable compromise with a straight line in terms of length of journey.

It could be added in support of the reverse curve that motion along such a path is generally experienced as pleasurable, that it is usually a visually attractive configuration and is an alignment which pervades nature from the meandering of rivers to the lines of motion of birds and fishes.

In terms of townscape it has similar advantages to those pointed out for a simple curve. Since speed and curvature are interrelated, so that the smaller the radius the lower the speed, it can be seen that for the townscape alignment a series of small-radius curves constituting a meandering line would be one method of keeping speeds within a desirable limit. As the scale of the curves increases so too will the design speed and as will be seen in Part II, The Flowing Alignment, the serpentine line in the form of large-radius reverse curves is ideal for that alignment.

2.48

The graceful line of the major access road, **2.48**, illustrates the attractiveness of these curves. The line of the road echoes that of the brick garden walls and is visibly dictated by the layout of the buildings. The visual logic is underlined by the sloping of the ground down to the left-hand side of the road so that it fits closely into the land form.

By reducing the scale and emphasizing the serpentine nature of the line the designer of the service vehicle access road in **2.49** has ensured very low speeds in a highly urban setting. The line of the meandering route is emphasized by the brick channel along the centre-line but a number of devices render the edges less precise than is commonly the case with access roads, thus inducing increased caution in the driver. These devices include carrying the brick surface past the bollards on the left-hand side and therefore past the running surface, not keeping the containing brick walls parallel to the centre line, using three different vertical edge elements in a short length of road, and changing the surfacing material from brick to bitumen macadam at a line which passes diagonally across the surface. A further inducement to slow speeds is the solidity of the containing elements which would inflict more damage on the vehicle than they would be likely to sustain in the event of a bump.

2.49

2.50

Variable width roads

With the flowing alignment it is obviously vital that the carriageways are a constant width; in other words, the lines of the carriageway edge must be parallel. This is not necessarily the case for the townscape alignment, with its different priorities and possibilities. Varying the width can be used to control speed, provide passing places and in special circumstances parking places, and to integrate the road more easily into the surrounding built environment or ground contours.

Variations in the road width can be incorporated in informal or formal road layouts. The casual relaxed appearance of the informal type warns the driver that he is in an area where slow movement is the norm, **2.51**. **2.50** shows the line of a minor access loop. This is a one lane, two-way road with frequent passing places where a hard surfacing is provided adjacent to the running surface in order to afford parking space and pedestrian access to the dwellings.

2.51

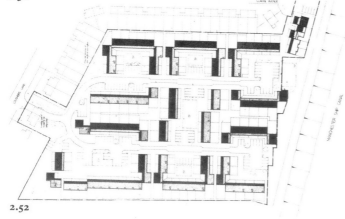

2.52

In the housing area of **2.51** it can be seen that in spite of the narrow, variable width road there is ample room for vehicles to manoeuvre.

In the formal layout of **2.52** the normal road width of 4.75 m is narrowed down over specific lengths to 2.75 m. This configuration will induce drivers to proceed at a low speed and at the same time they will have ample warning of the need to stop and allow the passage of oncoming cars or service vehicles.

The Brow at Runcorn, discussed in Chapter 1, Networks (see page 31) provides some other examples of the informal type of road, as does the layout of Murdishaw phase 1 at Runcorn New Town, Cheshire, 2.53 and 2.54.

The principles of variable width roads are beginning to be applied in some established urban areas. Where the road width supplied is in excess of that required for moving traffic and where it is desirable to keep through traffic out then that section of road can be narrowed at either end, 2.55. This slows vehicles, dissuades those whose destination is outside of that length of road and provides space for parked cars. An example of this alteration to an existing road is shown in 2.56. It is sometimes possible to stagger the entrance and exit. Another variation is the use of cantilevered barriers across the entrances.

Care should be taken in the design of variable width roads with two-way traffic to ensure that there is no ambiguity regarding the possibility of two vehicles passing at a particular section. They should also be arranged wherever possible to avoid the need for backing into passing spaces.

2.53

This layout is developed as a series of courts arranged between serpentine alignments of dwellings. The running surface is laid out in an informal manner, narrowing and widening to control movement and provide parking areas in carefully chosen locations close to the dwellings, 2.54.

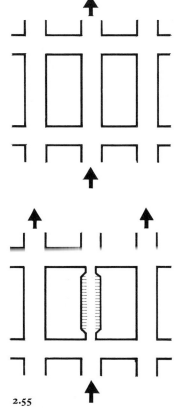

2.54

2.55

2.56

Loops

A loop road is a useful type of alignment when it is desirable to exclude through traffic without resorting to a cul-de-sac. To be effective the journey time for vehicles on the primary road should be significantly less than if they made a diversion through the loop. The journey time is the critical factor although this is usually directly related to the length. However although the loop could entail traversing a length of road considerably in excess of the equivalent length of primary road, it could still offer a quicker journey if the traffic on the main road is held up at traffic lights, roundabouts or in traffic jams.

Loops can be constructed on all types of roads except, of course, the main distributors and trunk roads. Since we are concerned here with townscape the smaller scale loop will be discussed.

There are a number of basic configurations, **2.57**. Loop roads can start and end at the same location on the primary road, (a).

They can also start and end on a different position on the same primary road, (b).

or start on one primary road and end on another, (c).

The scale on which loops can be used varies considerably even within the townscape context. An example of very small scale loops serving only a few dwellings is shown in **2.58**. These two loops belong to category (b) in relation to the access road. The example of **2.59** on the other hand is a much more complicated arrangement and belongs to category (c) as well as (b).

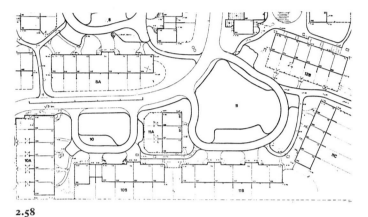

2.58

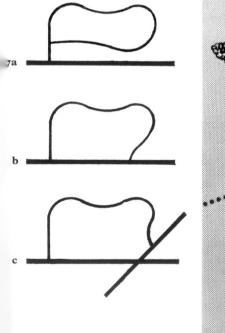

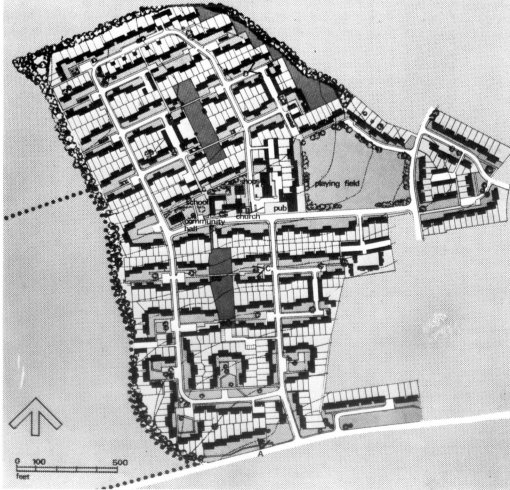

2.59

Loops can make the tasks of service vehicles easier than is the case with culs-de-sac, in which they must turn round and retrace their route. Also the often large areas of surfacing required for service vehicles to turn round at the end of culs-de-sac can be dispensed with. For this reason hard surfaces are sometimes provided linking the ends of culs-de-sac, these surfaces being provided for the use of service vehicles only (or service vehicles and pedestrians). A further advantage of loops over culs-de-sac is that since any part of the route can be approached from either end, the task of repairing the road or services under the road is more simple. Consequently a path linking culs-de-sac in the manner described is sometimes provided for use only in such an emergency, 2.60.

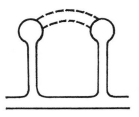

2.60

Culs-de-sac must obviously be two-way roads but as soon as two are joined at their ends to form a loop there is the possibility of introducing a one-way system. One-way loops have considerable advantages especially when one lane will cope with the total volume of traffic. It then becomes possible to consider the use of narrow width roads without passing places. In designing such a road for a residential area it should be remembered that provision should be made for service vehicles so that they do not block the road for private cars.

A solution to this problem may entail the provision of a limited number of areas where service vehicles such as ambulances, refuse vehicles and fire appliances can overtake or be overtaken by private cars. It is sometimes feasible to make an order exempting service vehicles from the one-way system at stated hours.

The possibility of cars breaking down or even being parked in a one-way system should also be borne in mind by the designer.

The observations made here on the subject of loop roads mainly relate to minor roads and the residential system. In other areas the use of one-way roads is essentially a technique of traffic management whose purpose is to increase the capacity of roads and thus does not come within the scope of this book.

A major benefit of one-way traffic accrues from the reduction in conflicting vehicle movements. The maximum capacity of one lane at the speed which maximizes the traffic volume is, for traffic in one direction, 800 vehicles per hour. If passing places and two-way traffic are introduced then the total capacity for two-directional traffic on one lane will be in the region of 150 to 200 vehicles per hour. At cross roads formed of two-way roads only there are 16 possible points of conflict. When one of the roads is one-way this figure reduces to 5 and if both the roads are one-way there is only 1 point of conflict, 2.61.

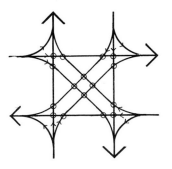

2.61a Crossroads with two two-way roads: 16 points of possible conflict

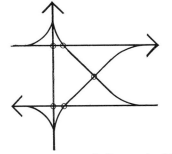

b Crossroads with one two-way road and one-way road: 5 points of possible conflict

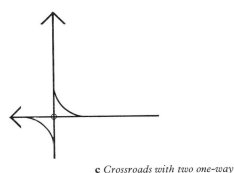

c Crossroads with two one-way roads: 1 point of possible conflict

Stopping sight lines can be less on a one-way single lane loop road than on a two-way single lane road or a two-way two lane road.

Long one-way loops pose some special problems. Drivers whose destinations are near the exit of a long loop may be tempted to drive along the road in the wrong direction hoping not to meet any oncoming traffic. Also drivers on journeys within the length of a long loop may feel disposed to risk driving against the flow rather than undertake a long detour. It is usually best to keep the loop short enough to avoid these problems becoming serious.

The designers should bear in mind that pedestrians when crossing a road expect the traffic to be approaching from the right. Where specific pedestrian crossing areas are provided in a one-way or a one lane two-way system consideration should therefore be given to the provision of a warning about the traffic directions. One-way roads generally are likely to require more signing than two way systems and the implications of this in terms of the effect of signs on the quality of the environment should also be taken into account when debating the use of a one way system.

Culs-de-sac

The cul-de-sac is a traditional method of eliminating through traffic and creating quiet precincts in housing neighbourhoods. In these areas it ensures that the access road will be used only by cars and service vehicles whose destination is a dwelling served by the road. The benefits of this are clear. The study by Noble and Adams, *Housing : the home in its setting*[1] remarks that 'The exclusion of through traffic from the immediate housing area is the most useful precaution which can be taken to reduce danger to pedestrians, unnecessary noise, vibration, dirt and fumes'. Although housing areas should always include adequate outdoor space for children's play it is never possible to keep them entirely off the access roads, so that even in schemes where childrens' needs are catered for elsewhere it is necessary to make the roads safe for them. On this subject the DOE publication comments 'It will in many cases be possible to combine convenient vehicle access to the dwelling with reasonable safety for children and pedestrians. Our surveys suggest that some short access roads and culs-de-sac are considered by parents to be reasonably safe for play'.

A cul-de-sac may be a simple short spur off a through road or it may branch out into a complex system, **2.62**.

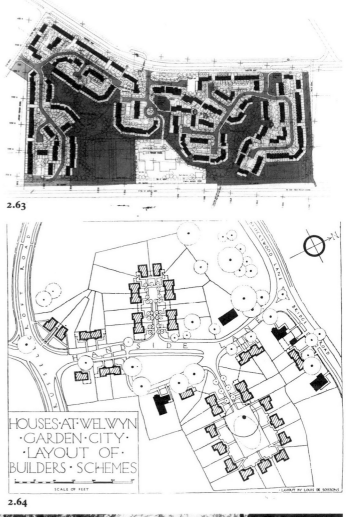

2.63

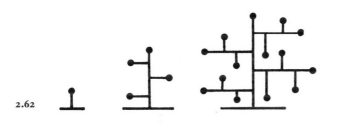

2.62

HOUSES·AT·WELWYN
·GARDEN·CITY·
·LAYOUT·OF·
BUILDERS·SCHEMES

SCALE OF FEET

2.64

.65

An example of the first type is depicted in **2.63**, a residential development in Welwyn Garden City, Herts. Photograph **2.64** shows the entrance to a typical short cul-de-sac in an older part of the New Town. This access serves a small number of dwellings and ends at the building seen on the road centre line in the photograph.

The more complicated arrangement of **2.65** is at Thamesmead Area 10, Kent. This plan shows one way in which the use of culs-de-sac can be combined with the provision of large soft-landscaped areas and play areas to afford rival attractions for children which will help to keep them away from the roads. The serpentine alignment assists the designer in developing varied and interesting spaces as well as controlling vehicle speed.

In the traditional cul-de-sac the dwellings faced onto the road and pedestrian access to the front of the house was along footpaths adjacent to the road, **2.66a**.

The Radburn type of layout was conceived shortly after the First World War by the Regional Planning Association of America. The purpose of the layout was to separate pedestrians from vehicles in the residential areas of towns. The Radburn plan involved the use of a series of superblocks of 5 to 10 acres, each one being surrounded by a radial road with the enclosed area penetrated by a series of culs-de-sac. No road went right through this area. It was possible to walk throughout the superblock without crossing a road and the blocks were interconnected by a system of pedestrian underpasses as well as roads. The arrangement of a superblock is represented diagrammatically in **2.67**. The dotted lines indicate footpaths leading to some focal area within the radial road.

In the original development at Radburn the dwellings faced the culs-de-sac and the pedestrian ways passed between the rows of back gardens. Subsequent developments were arranged so that the dwellings overlooked pedestrian precincts or corridors whilst the culs-de-sac provided vehicular access to the backs of the houses **2.66b**. More recently in this country the ends of the culs-de-sac have been positioned to provide easy access to segregated footpath systems, **2.66c**, and often provision is made for pedestrians also to follow the lines of the roads.

The road corridor for culs-de-sac should be designed in accordance with good townscape principles as with any other type of road. The paved surface at the end of the culs-de-sac will have to be widened in order to accommodate turning vehicles. There are many ways in which the end space may be treated. The most successful of these integrate the turning facility into the total design.

Turning vehicles must be provided with a certain minimum surface area on which to manoeuvre. It has been common practice in the past to limit the area provided strictly to the theoretical minimum required. This can result in a rigid geometric layout, unsympathetic to the needs of the local area, **2.68a**. It is often a simple matter to soften the edges by introducing more organic looking curves, **2.68b**, which can be fitted to local ground contours and planting layout. An example of a relaxed treatment which provides ample turning space is shown in **2.69**.

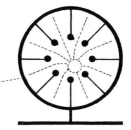

2.67

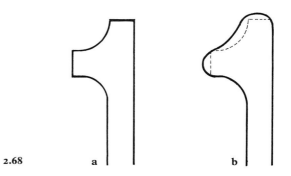

2.68   a   b

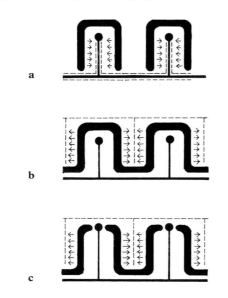

2.66   c

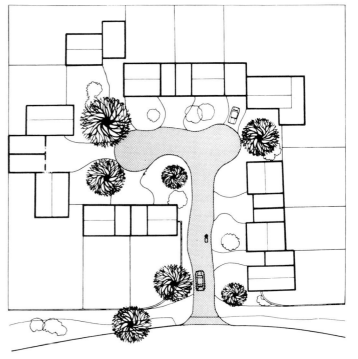

2.69

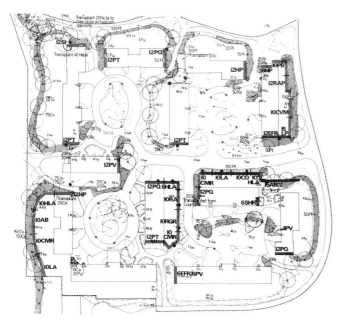

Another method of treating this terminal area is to incorporate a small scale loop, thus avoiding the need for vehicles to use reversing movements in turning. This is the method which has been used at Longbarn 4, a housing development in Warrington New Town, Cheshire, **2.70**. The end of the cul-de-sac on the bottom right-hand side of **2.70** shows the use of a turning and parking area with curved, free flowing edges. A photograph of part of a typical turning area is shown in **2.71**.

71

The realization that the greatly increased numbers of vehicles on the roads in recent years is having seriously detrimental effects on the environment in many established residential areas has led to attempts to modify existing layouts to alleviate these effects.

One such attempt involves the conversion of existing through roads to culs-de-sac. For example in a residential area laid out in a grid pattern it is sometimes advantageous to make part of a road element into a pedestrian-only area, thereby converting the through road into two culs-de-sac, **2.72**. This can provide communal play space which is often lacking, and a sitting out area, at the same time stopping through traffic and providing parking space for the local residents.

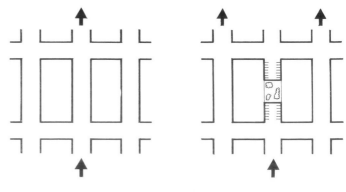

2.72

65

Squares

Squares are a delightful way of providing outdoor landscaped space within a formally planned residential development. They are created by designating an area within a grid network of roads for use as communal open space for the benefit of local residents or the public at large. In its simplest form it consists of one of the modules of a square grid which is landscaped, the module being surrounded by others containing buildings which enclose the space, **2.73a** and **b**. The creation of such an area introduces a breathing space within the urban fabric. It adds to the variety of views along the roads which pass the perimeter of the open space and provides pleasant outlooks for the inhabitants of surrounding dwellings as well as establishing a pleasant site for leisure use.

The more visually contained the space, the better it will contrast with the linear spaces of the roads and the greater is likely to be the sense of arriving at a unique place within the routes of the town or city.

Squares provide a site within an urban context where forest trees can develop fully. In many of the London squares, some of which are illustrated here, full advantage has been taken of this opportunity. These trees modify and complicate the overall volume enclosed by the surrounding buildings. The perimeter roads are contained within their own space, being confined on one side by the buildings and on the other by the tree canopies. And the canopies define the ceiling of the larger space of which the roads are also a part.

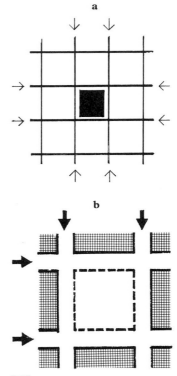

2.73

The arrangement shown in **2.73** is not ideal from the standpoint of containment, since the views in the direction of the arrows are not arrested by the presence of the square. Similarly the person in the square has an unhindered view out. Nevertheless many squares of this type have been built. A modified example with two squares arranged in alternate blocks is shown in **2.74**. Here some enclosure of views has been achieved by the use of T-junctions.

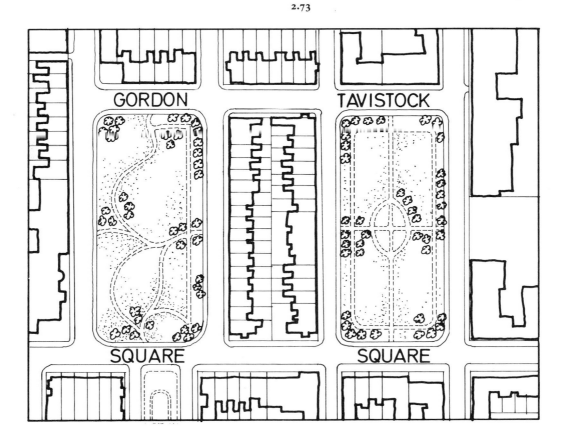

2.74

66

There are many methods of ensuring that the views into the square and those out along the line of the perimeter road are curtailed. Some of these are outlined below.

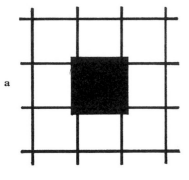

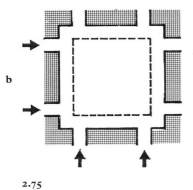

2.75

Method 1
If the size of the square is greater than the area defined by the road grid, **2.75a** and **b**, then the views will be blocked.

This solution has been used in the design of Belgrave Square, Westminster, **2.76**. When the roads do not run straight through an architectural problem arises in the layout of the buildings at the corners. In this case it has been solved by placing a building across the corner at 45° to the sides of the square, **2.77**.

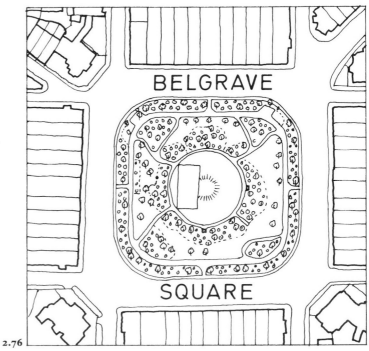

BELGRAVE

SQUARE

2.76

.77

67

Method 2

When the perimeter road terminates at a T-junction a short distance from the square the sense of containment will be improved, **2.78a** and **b**. Also the square will have a more secluded character than would be the case if it were bounded by main through roads. A partial example of this method occurs at King Square, Islington, London, where President Street terminates in a T-junction with Goswell Road, **2.79**.

Method 3

This method entails the omission of part of the grid in the vicinity of the square so that T-junctions are formed at its corners. Figures **2.80a** and **b** show a theoretical example in which the road along each side of the square terminates at one end in a T-junction at the square and at the other end in a T-junction one block distant from the square. A partial example is shown in **2.81**, where Sterling Street along one side of Montpelier Square, Kensington, London, is of this type.

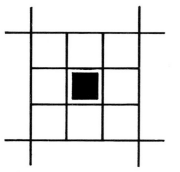

2.78a

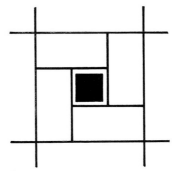

2.80a

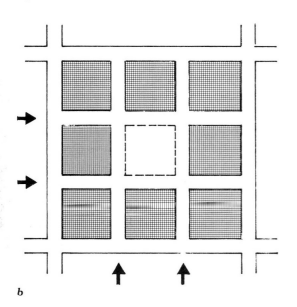

b

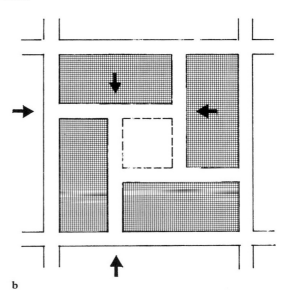

b

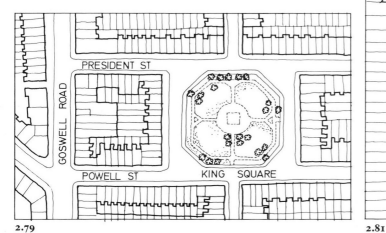

2.79

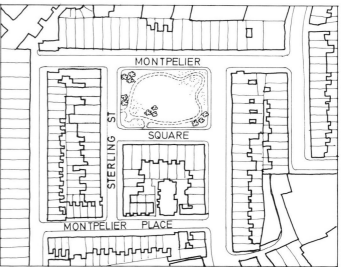

2.81

Method 4

If alternate rows of blocks running say west-east in a square grid are displaced a distance half a block in length, then the north-south routes will form T-junctions with those at right angles, **2.82a** and **b**. Thus the views along the north–south routes are interrupted and a partial solution of the containment problem results. In the London squares planted with large trees the vistas along these approach roads are terminated by a wall of vegetation, a welcome relief from the hard containing surfaces elsewhere. This is the type of layout to be found at Bloomsbury Square in the London Borough of Camden, **2.83**.

Photograph **2.84** shows a view looking along Great Russell Street towards Southampton Row. The effect of the T-junction combined with the tree canopy in forming a particular space containing the road is apparent. (The designers' emphasis on the static rather than the dynamic qualities of the space seems to have been successfully conveyed to its occupants.)

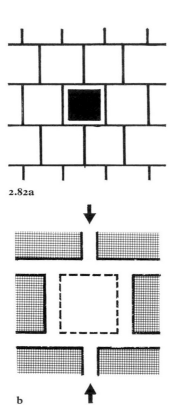

2.82a

b

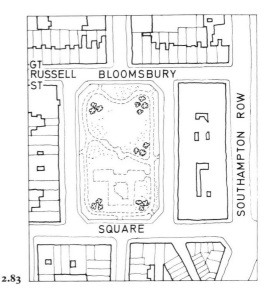

2.83

2.84

Method 5
More complete containment
results when the square straddles
the intersection of two routes at
right angles, **2.85a** and **b**. An
example of this solution is that of
Soho Square in the London
Borough of Westminster, **2.86**.

Method 6
A variation on the theme of
Method 5 is achieved when the
square straddles an intersection
but is turned at 45° to the axis of
the grid. This is equally effective
in blocking the views, yet the
layout has certain characteristics
which are different from those of
the previous type. Notably the
intersection of the incoming road
with the perimeter road becomes
a Y rather than a T-junction and
the structure on the corner must
negotiate a 45° change in
direction rather than one of 90°,
2.87a and **b**. This type of solution
is illustrated by Northampton
Square in the London Borough of
Camden, **2.88**.

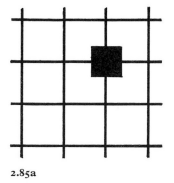

2.85a

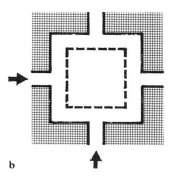

b

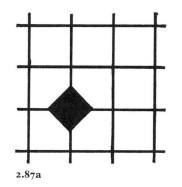

2.87a

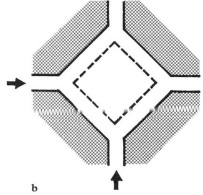

b

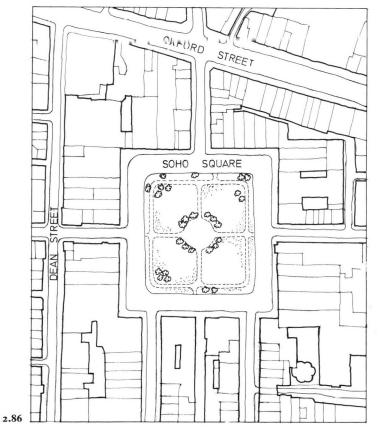

2.86

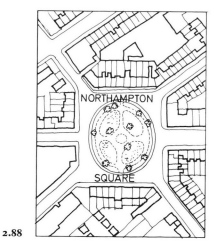

2.88

Method 7

A discontinuity in the grid involving the termination of the elements in one direction before the grid square is reached is another way of making the square more secluded **2.89a** and **b**. Illustration **2.90**, showing the plan of Ovington Square in the London Borough of Westminster, is a case in point.

Method 8

This is a variation on Method 7 in which the square is situated on one side of a through road with no access except from this road. In effect the perimeter route round the square becomes a loop off the through road, **2.91a** and **b**. Brompton Square in the London Borough of Westminster illustrates this type, **2.92**. Clearly a high degree of seclusion is enjoyed by the residents who overlook this square since through traffic is completely excluded.

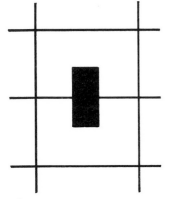

2.89a

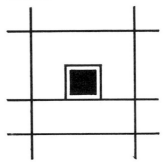

2.91a

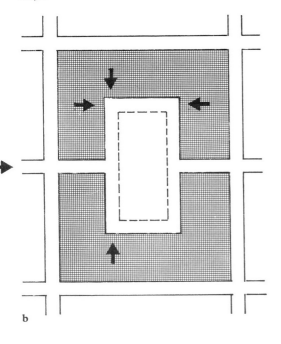

b

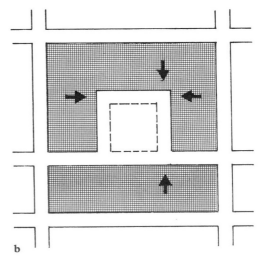

b

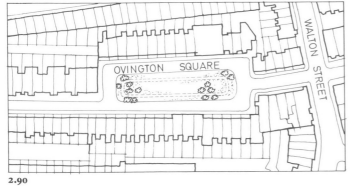

2.90

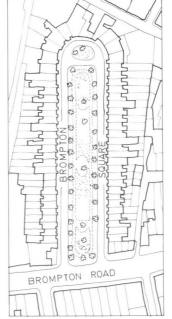

2.92

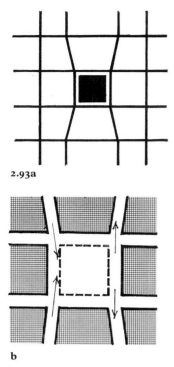

2.93a

b

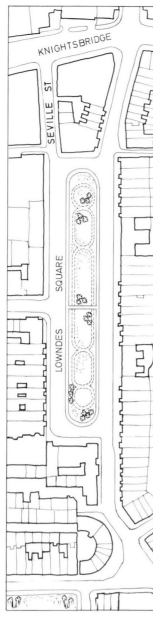

2.94

Method 9
A change in the direction of the grid line as it leaves the square is equally effective in attaining the required end, **2.93a** and **b**. An example is Loundes Square in the London Borough of Camden, **2.94**. Even an apparently very slight change in direction as seen in plan, such as that of Seville Street, is highly effective in closing the view when seen on the ground.

In the traditional London square a perimeter road usually separates the buildings bounding the space from a soft-landscaped central area of the same plan shape as the square itself. This is not invariably the case and a few variations are mentioned below. Normally London squares, as well as being grassed and containing trees and shrubs, are surrounded by a railing or fence; access points are thus restricted. Sloane Square, **2.95**, is a highly successful departure from this norm, being of a type more frequently found on the continent than in this country. The central area is hard-landscaped with a fountain as the focal point, and the surface is shaded by the canopy of two rows of London plane trees, **2.96**.

2.95

2.96

It is, of course, possible to design
layouts in which the perimeter
road does not go round all the
sides of the square. In the older
areas of London such an
arrangement occurs but tends to
be called by some other name.
Ennismore Gardens, Kensington,
is arranged in the form of a
square but with a perimeter road
on three sides only, **2.97** and
2.98a. Photograph **2.99** gives an
impression of the character of this
area. It would be equally possible
to design a layout with the road
on two sides or one side, the final
stage of seclusion in relation to
vehicles being those squares
which are enclosed by structure to
the exclusion of all but pedestrian
traffic, **2.98b** and **c**.

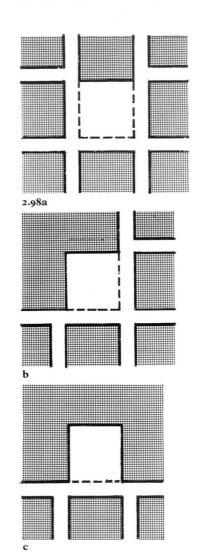

2.98a

b

c

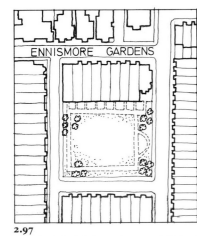

2.97

.99

73

2.100a *Simple crescent*

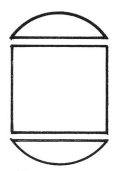

b *Crescents terminating a road*

c *Crescents in association with square*

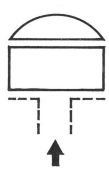

d *Crescent in association with twin squares*

e *Crescent terminating vista through park*

f *Compound curve crescent overlooking square*

Crescents

A crescent is a row of buildings laid out on a curved line; usually the segment of a circle. This form has often but by no means always been used as an element in a formal geometric arrangement of buildings.

A few ways in which crescents have been used in the past are outlined below. These layouts can be represented by the thumbnail sketches of **2.100.**

The simplest use of a crescent is as a loop off another road without the use of any other special townscape elements, **2.10a.** The area between the crescent and the primary road may be used for building or, as is often the case, designed as a communal or public garden, **2.101b.**

The crescent may start and end at the same road without any other access being provided. This has the advantage of maximizing the amount of privacy. Alternatively a road may enter the crescent some way along its length, commonly at its mid point, **2.101c.** If gardens are used they will terminate the view of a person approaching the crescent along this road. The person will not be aware of the existence of the crescent until he reaches the road end. He will then be faced with the choice of turning to the left or right and in each direction the dramatic curve of the building elevation will suddenly become apparent. This configuration then provides a powerful townscape arrangement.

2.101a *Simple crescent with island segment built upon*

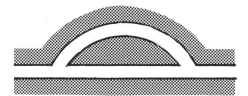

b *Simple crescent with island segment used as garden*

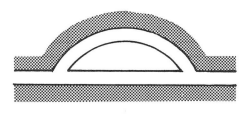

c *Simple crescent with island segment used as garden and additional incoming road*

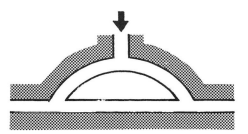

74

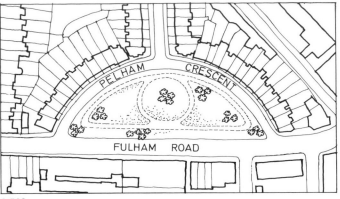

2.102

2.103

2.104

Pelham Crescent in the London Borough of Kensington and Chelsea, **2.102**, provides an example of this type, with an incoming road and a communal garden. A view of an entrance to this crescent from Fulham Road is shown in **2.103**.

Crescents provide an imposing termination to a road, **2.104**. When the island segment is in the form of a garden then, as the end of the road is approached, the vista presented by the crescent rapidly unfolds. Alfred Place in the London Borough of Camden, **2.105**, is an example of this plan arrangement, although the full potential of the layout is not realized due to a structure having been sited on the island of the North Crescent.

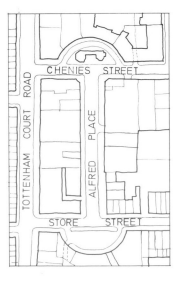

2.105

Crescents can be used in association with squares, **2.106**, either with or without buildings on the crescent island. This use of crescents is one way of containing the views out from the square. When an incoming road is located within the crescent, as in **2.107**, then the crescent provides a gradual transition from the road to the square.

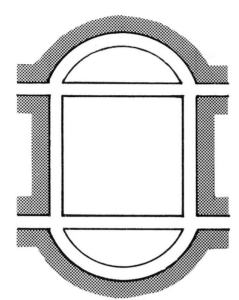

2.106

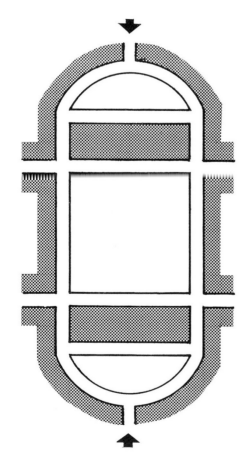

2.107

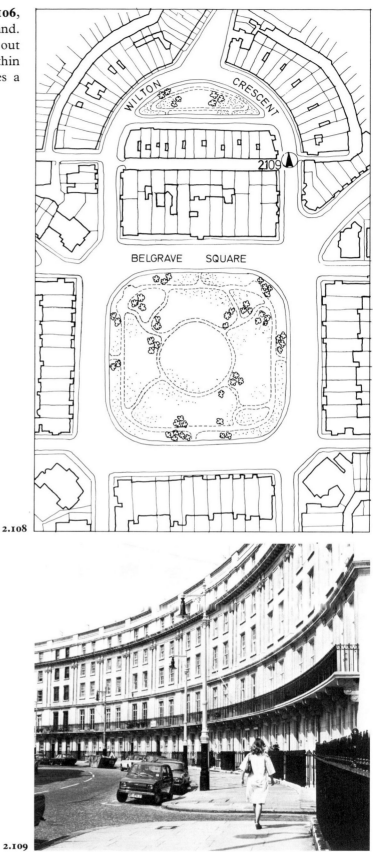

2.108

2.109

By locating a block of buildings between the crescent and the square, separate contained spaces of contrasting shape are created which are experienced in series by the person moving through the complex, **2.107**. This is the arrangement which has been adopted on the north side of Belgrave Square in the London Borough of Westminster, **2.108**. A photograph of Wilton Crescent is shown in **2.109**, the position being arrowed on the plan, **2.108**. From this view the canopy of trees can be glimpsed and, as the crescent is entered, the gardens become progressively more dominant.

A more complicated arrangement of crescents and squares is seen in plan **2.111**. This layout results in a number of interconnected spaces of different character. St John's Church has been thrust forward into Hyde Park Crescent so that it blocks the view BF, **2.110**, through the crescent and becomes the focal point for all incoming routes at B, C, D, E and F as well as for all views from buildings along the crescent.

The squares have been arranged in relation to the crescent, in this case, so that views A and G are contained by the buildings lining the roads at B and F rather than leading directly into the crescent. By using a curved road at C the designer has ensured that the church will appear quite suddenly to view whereas it can be seen from some distance away as the culmination of the other routes.

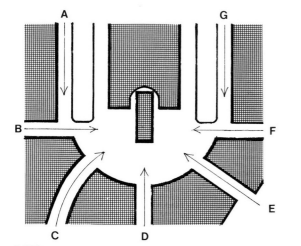

2.110

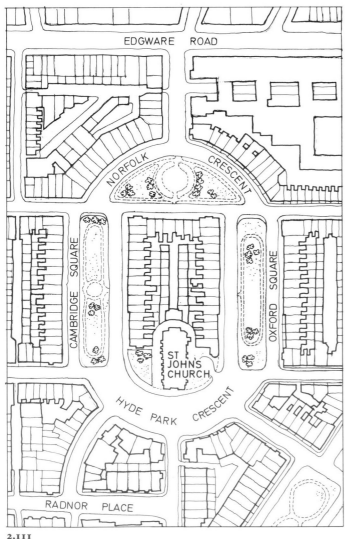

2.111

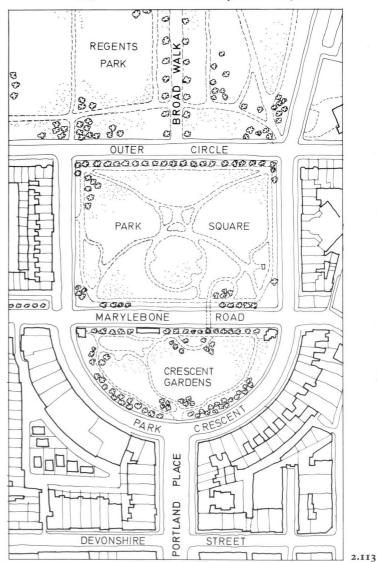

2.112

Crescents can be used to good effect in relation to much larger open spaces. In the layout of **2.113**, Park Square and Park Crescent in Regent's Park in the London Borough of Camden, are the climax of Broad Walk for those walking south through the park towards Portland Place. By penetrating the urban fabric this area forms a transition between the parkland and the hard landscape of the city.

The use of a crescent in relation to the park provides a more gradual transition from the open space into Portland Place than would have been the case if the road had formed a T-junction with Marylebone Road, **2.112**.

2.113

Although a crescent is often in the form of a segment of a circle, other alignments can be and sometimes are used. A portion of an ellipse has been adopted and occasionally a compound curve. These more complicated arrangements can introduce difficult problems of layout and detailing. In a terrace of dwellings for example, the problem of obtaining a sufficient standardization of floor plans could be considerable. Nevertheless the townscape characteristics of more complicated layouts can make their use justifiable in some circumstances.

Avenues

One of the meanings of this term is a wide road, often planted with trees in regular rows, usually straight and formal in appearance. Designed to impress the viewer, avenues are important features in many cities and towns.

2.114

2.115

A well known example is Pall Mall in Westminster, London, leading from the Admiralty Arch at Trafalgar Square to Buckingham Palace. As can be seen in photograph **2.114** the route is straight, and perfectly aligned with the Palace. It focuses on the Victoria Monument, which arrests the traffic movement before the palace is reached. The monument is situated on a roundabout and this disperses the traffic to the roads on either side of the Palace.

The sense of grandeur and an awareness that this is a ceremonial way are clearly conveyed by the proportions of the design. The space allocated to pedestrians is as generous as that provided for vehicular traffic, **2.115**.

Avenues need not always be designed to such a grand scale. They are also, for example, incorporated in residential developments and used as approaches to large single dwellings. But even though the scale may vary they always maintain their essentially formal and imposing character.

Courts

The use of courtyards as contained external space adjacent to single buildings, and courts as spaces enclosed by a group of dwellings are other traditional forms of access which can be part of the repertoire of today's designers.

Squires Mount in Hampstead, London, is an elegant example of a court serving a small group of dwellings, **2.116**. The sign 'private' on the gatepost, **2.117**, is almost superfluous, as the gateway itself eloquently conveys this message. The space within has a powerful sense of enclosure created partly by the height of the buildings in relation to the width of the court, the mass of foliage in the background and the ceiling provided by the canopy of trees. The atmosphere of privacy pervading such a space is similar to that of a private living room. The feeling of intimacy is greatly enhanced by the meticulous detailing both in respect of the materials chosen and the way they are used. The rounded stones set in mortar which constitute the main surfacing, filling the panels delineated by the lines of brick paving, provide strong texture. Where rough textures are employed in dual use areas the incorporation of paths of smooth-surfaced materials along pedestrian routes can make circulation on foot more pleasant. However in this case the main purposes of the brickwork are to create a pattern and reduce the scale. A practical effect of the use of such a rough surface is that oil stains will be much less conspicuous than would otherwise be the case.

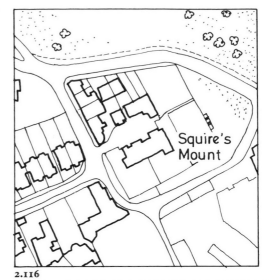

2.116

117

Major contributions to the character of the space are made by the irregular building line, the well-proportioned elevations and the fact that the surfacing runs right up to the face of the building.

The close proximity of the planting to the hard surface creates an intimate relation between the hard and soft materials, heightening the impression of exclusiveness. It is noteworthy that cars look as much in place in such a space as do pedestrians.

At Tree Close in Petersham, Surrey, a square dual use court forms the central open space, enclosed by the twelve single-storey houses comprising the scheme. The houses are arranged in four clusters of three, each cluster enclosing a small pedestrian courtyard, **2.118**.

The curved entrance to the court imbues in the motorist and pedestrian alike a sense of arrival when the bend is negotiated and the central space is reached. It also has the effect of increasing the privacy of the space. The wide band of brickwork at the road edge is a sensitive substitute for a kerb; it marks the road edge

without interrupting the pleasant flatness of the land and thus the road appears as an integral part of the landscape, **2.119**.

The large area of in-situ concrete has been broken up into smaller areas by the grid of insert brickwork, thus an appropriately small scale is achieved, yet such an area allows even the largest service vehicles to manoeuvre with ease without the need to incorporate turning heads, **2.120**. The change from asphaltic road materials on the access road to concrete in the court and the concurrent change from a curved line to a square area assist in warning the motorist that he is entering a space where special caution is required and the rights of the pedestrian are dominant. As the trees mature the sense of enclosure will be increased and shadows cast will soften the effect of the in-situ concrete.

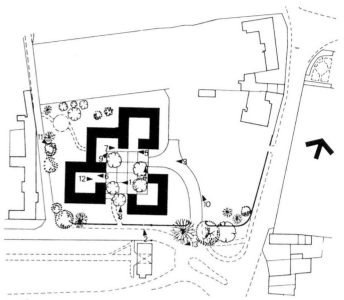

2.118

2.119

80

2.120

Oxclose Village in Washington New Town utilizes the court principle on a larger scale than the previous example. This is one of 18 separate and self-contained village communities which will eventually constitute the New Town.

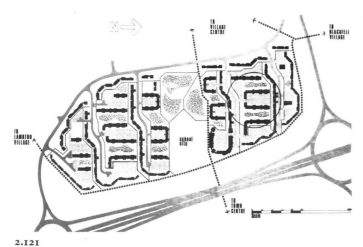

2.121

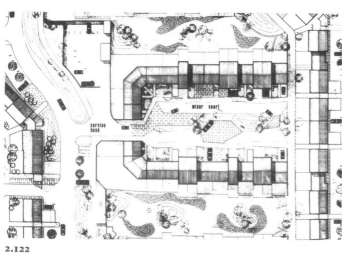

2.122

Housing Projects 1, 2 and 7 occupy the eastern sector of the village. As can be seen on the general layout plan, **2.121**, the dwellings are set out in groups, each group enclosing a court. These courts are designed to accommodate cars as well as pedestrians but to exclude service vehicles. The latter are only permitted to penetrate to the short spurs off the culs-de-sac access roads, as shown by the area marked 'service head' on **2.122**. The access roads join a distributor road which links directly with the town's main highway system.

Partial segregation of pedestrians and cars has been implemented in the layout. The pedestrian routes are shown by dotted lines on the general layout plan, **2.121**. The enclosed spaces are called 'mixer courts' by the development corporation, and each mixer court forms a bridge between the main pedestrian walkways and the access roads. Through their form and surface materials they are designed to inhibit the motorist and give priority to the pedestrian. The mixer courts are laid out to alternate with semi-enclosed public gardens on the side of the dwellings to which there is no vehicle access.

81

Mews

Although the mews takes different forms in different parts of the country it is essentially a small-scale, enclosed space providing primary or secondary access to dwellings, the total area of the ground surface being used by both pedestrians and vehicles. The space is usually contained by a small group of dwellings with the dual use surface extending to the building line.

This form of spacial arrangement offers many opportunities for the designer to create interesting townscape. A brief look at some existing mews will provide an indication of the wide spectrum of possibilities which exist within this one type of solution to the problem of providing access to dwellings for pedestrians and vehicles. Particular attention is drawn to the traditional London mews only because it contains a rich source of material with which to illustrate some features of this access form. Mews in other cities and towns could equally well be selected for study.

The London mews were initially formed as service areas at the backs of properties. In the example illustrated below those parts of the properties facing onto the service area have been converted to houses in their own right and so the mews has become the main access to a cluster of dwellings.

2.123

2.125

2.124

Eaton Mews South in the London Borough of Westminster is a typical example of this type of development. When viewed from along the length of the road the entrance to the mews is inconspicuous and does not significantly interrupt the continuity of the building line, 2.123. It is narrow, discouraging casual intruders looking for a parking space, and is clearly proclaimed by the imposing archway. This archway defines one side of the simple volume of which the dual use surface forms the floor. The view of photograph 2.123 clearly conveys the impression of the mews as a semi-private space inside the arch in contrast to the public space outside. The slight ramp down at the entrance reinforces the sense of 'going in' and the narrow entrance, the archway and the ramp dispel any ambiguity there may have been as to whether this constitutes a through route. (The design of the street sign, although logical, can hardly be considered pleasing. A distinct Heath-Robinson impression is conveyed by the widened base, the narrow shaft, the clip-on rectangular 'one-way' plate, the disc containing the arrow symbol, all topped with the obviously added small lamp. The design of street furniture is a matter which requires much more attention than it is accorded. Even our most attractive streetscapes are commonly polluted by poor designs.)

Inside the arch there is no distinction between the domain of the vehicle and that of the pedestrian. This in itself inspires caution in the driver which is further encouraged by the rough texture of the surface, the sense of confinement conveyed by the large height/width ratio and the obvious absence of other entrances and exits, 2.124. Variations in the building line make structural volumes apparent and complicate the volumes of the external spaces, resulting in a general heightening of interest including the attraction of a more intricate pattern of light and shadow than would otherwise be the case, 2.125. The recesses and alcoves formed can be used as parking places and as discrete positions for doors and stairs into dwellings.

The view in photograph 2.126 emphasizes the powerful impact of the archway in conveying the impression that the mews is 'inside' and is a private 'place' whereas the road is outside and a public 'route'.

In spite of the narrowness of the entrance it does not appear that the pedestrians on the footpath would be in any danger as they walk across the opening.

The distance between the massive kerbs is rather less than the distance between the legs of the archway. This has two advantages; a pedestrian can stop and look into the mews before alighting from the kerb and the arch is protected from vehicles striking the archway supports as they pass through.

2.126

The mews continues to be a highly desirable form of layout in many circumstances. Two examples of its use in recently constructed developments are presented below.

2.127

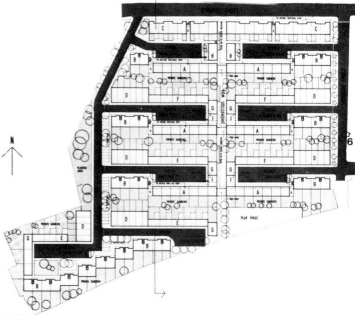

2.129

2.128

2.130

2.131

Although the houses usually overlook the mews court it may be contained by either the front or the back of the dwellings. A modern example at Coronation Street, Cambridge, has back gardens adjoining the mews on one side. The ends of these can be seen on the left-hand side of photograph **2.127**.

Integral garages have been included in some of the dwellings. The arrangement used, with entrance from the back garden, has the advantage that the car owner has the option of parking his car within the curtilage of his dwelling even if he does not wish to garage it. This can be an important consideration in areas where vehicles are subject to vandalism.

Photograph **2.128** shows the side of the mews opposite the ramp (seen in **2.127**). A through route has been provided for pedestrians but not for vehicles.

A recently built housing development at Nailour Street in the London Borough of Islington adopts the mews court arrangement in modified form. The site layout is reproduced in **2.129**.

Vehicles are constrained to slow down by the narrowing at the entrance and the steep ramp up to the mews court area, **2.130**. The height to width ratio is lower than that of the traditional London mews and a precast concrete upstand kerb has been incorporated with a raised footpath around the contained area. Integral garages have been provided and the external stairway up to the gallery entrances has an interesting equivalent in the older type of development. Compare **2.131** with **2.125**, page 82.

Further evidence of the continuing popularity of mews courts is provided by some of the design guides for the layout of residential areas published by county authorities and others. These are produced to encourage good design of residential developments and in particular designers are urged to develop layouts which are in character with the particular area in which the new building occurs.

A number of these guides recommend that the use of mews courts should be considered. As examples, diagrams of layouts produced by Essex County Council and Norwich City Council are shown in **2.132** and **2.133**.

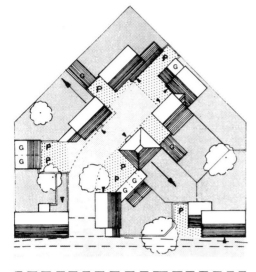

2.132

Pedestrian / Cycle Track

Access Road

P	Parking
G	Garage
▲	Front door
→	Main prospect
═	2m wall
⌐ ¬	Minimum highway area required

Private zone

Public zone

Adopted highway in public zone

2.133

The Essex Design Guide was the first of its type to be produced and some residential developments which were prepared in accordance with the principles laid down in it have now been completed. Photograph **2.134** shows a mews court in such a scheme. It can be seen that a well contained dual use space has been achieved.

2.134

85

Lanes

A lane is a narrow route for the passage of pedestrians and sometimes also vehicles. They are found in the countryside, in villages, towns and cities, and people are always intrigued by these confined, serpentine and unpredictable linear ways. It would be a pity if our current preoccupation with the needs of cars and large service vehicles were to mean that lanes were no longer considered suitable elements for urban projects.

Part of the charm of lanes is the way in which they arouse feelings of curiosity, anticipation and surprise, and of being protected from the elements by flanking buildings, walls, embankments and hedgerows. There are many ways in which this form of route can add to the variety of our surroundings but space will only permit the inclusion of one example.

2.136

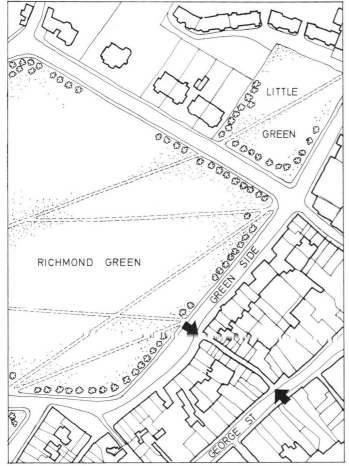

2.135

2.137

The small scale character of lanes, which people have always enjoyed, is most effective when contrasted with space of a different scale. An extreme example is the well known and popular area called The Lanes in Brighton which exists in dramatic juxtaposition to the great vistas of the sea front.

Another example is Brewers Lane in Richmond, Surrey which connects the main shopping centre of George Street with the village green, **2.135**. The sequence of contrasting spaces is illustrated in photographs **2.136**, **2.137** and **2.138**.

2.138

86

Alleys

These, like lanes, are traditional types of minor routes. Usually even narrower than lanes, the sense of being in a restricted linear space is frequently very strong. They are used as minor connecting ways between larger spaces.

.139

Even alleys can be designed to take service vehicles. The one shown in **2.139** connects a mainly residential street with the busy shopping thoroughfare of Bond Street in London. Such sudden changes in scale and atmosphere are one of the delights of the urban environment.

Village roads

Thomas Sharp, in *The Anatomy of the Village*,[5] points out that there are two main types of traditional English village, the 'roadside' type and the 'squared' type. The roadside type is essentially linear, having formed along an existing road or sometimes at a junction of roads, when the village extends along the legs of the intersection for a short distance, **2.140a**. On the other hand in the squared type of village the buildings are grouped round, and contain, an open space, **2.140b**.

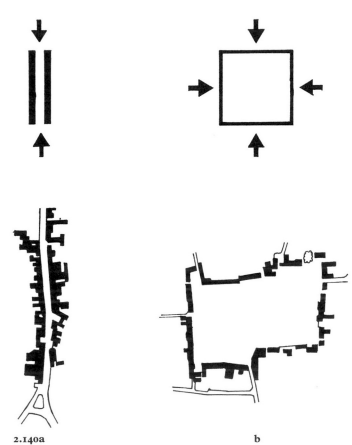

2.140a   b

Superficially these two types of village could be related to the concept of the environment being experienced as a series of routes and places as described at the beginning of Chapter 1, Networks. Yet it is one of the great attractions of traditional villages of both types that they are always places; highly individual places moreover with their own unique character.

The roadside village had to reconcile these two attributes, that it was constructed along a route and that it should be a recognizable place. The ways in which this has been done are a subject of special interest to all those concerned with the townscape alignment of roads. Sharp presents many village plans in his book, which repay detailed study. A few of these plans are reproduced here. In recommending a study of village alignment a slavish imitation of these is not being advocated. However an understanding of the way in which they work will provide valuable clues to the solution of some modern townscape problems.

There are many ways in which the variations in the route line contribute to making the linear village a place with an inside and an outside. One of these variations is a simple widening of the road at the village. This is often accompanied by a slight shifting of the axis, **2.141**. The sense of place is enhanced by the restricted size of the village and by the length of the village road. It is almost never possible to see through the village and out the other side by looking along the road. The views in and also the views out are contained. This is sometimes accomplished by slight changes in the direction of the route line.

2.144

2.141a   **b**

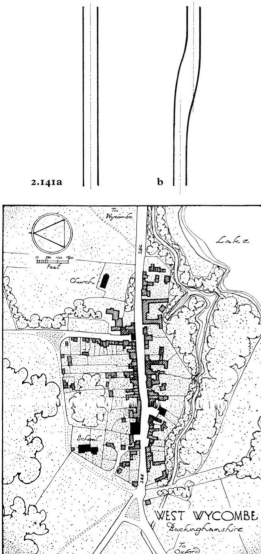

2.142

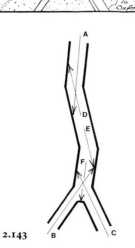

2.143

West Wycombe in Buckinghamshire is an example of a roadside village of the simplest form, **2.142**. In diagram **2.143** the shape has been exaggerated in order to illustrate the containment of the views. The view in at A is closed by the changes in direction of the road and the views in at B and C are contained by the fork. Similarly the views out at D, E and F are also contained. As can be seen on the plan, the changes in direction in the line of the road along the length of the village are very subtle. It is noteworthy that what appears at first on plan to be an almost imperceptible change in direction can in reality be completely effective in achieving the containment.

Another factor which sometimes blocks views through the village is the location of a particular structure or structures which appear to have been thrust out of line into the roadway for this purpose or to make the building more prominent. Often the building which is out of line is one of the more important structures for the village community. In the case of Coxwold in Yorkshire it is the church, **2.144**. By the means described, the view in at A, **2.145**, and the view out at B are blocked. At the opposite end of the road the end building of a row performs the same function, C.

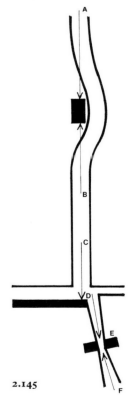

2.145

2.146

Another common factor
restricting views is the narrowing
of the road with buildings on
either side at the narrowed
portion. This is illustrated by the
constriction at the eastern
entrance to Coxwold village,
where E restricts the views out
and in at D and F.

The church in Thaxted, Essex,
interrupts the line of the road in
an even more definite manner,
2.146 and **2.147**, so that it is
exactly in line with the view of
the road user facing north west,
D, and is aligned with the view of
the person approaching from the
west, A.

The bifurcation of the road and
the construction of buildings on
the resulting island affects the
containment of the views F and
E. Furthermore the road
widening, the island site, and the
irregular road width produce a
number of diverse and visually
interesting spaces within the basic
linear form of the village.

Once more, the curves in the
approach roads and their
connection with the main village
road at an angle curtail all the
views in and out of the village. At
the north-west end the sense of
entering a place is heightened
further by the narrowing of the
road at B, just before the church
comes into full view, by means of
the proximity of two buildings.

2.147

The squared village

As Thomas Sharp points out, a squared village may be one
of many shapes including rectangular, triangular, and
irregular. The examples below indicate the ways in which
the entrances are contrived to close all the views in and out
so that the square is experienced as a protected, enclosed
environment.

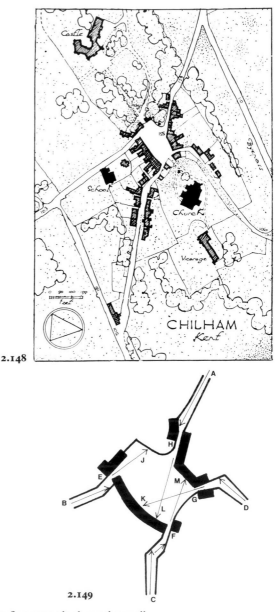

2.148

2.149

The first example shows the small
square of Chilham village in
Kent, **2.148** and **2.149**. All the
approaches, A, B, C and D, have
a kink in their alignment making
it impossible to see directly into
the square until it is reached. The
threshold to the square is
emphasized at these four
entrances E, F, G and H, by a
narrowing of the road. This
narrowing is effected in each case
by the proximity of two buildings
on either side of the access road.
Also in every case the view into
the square is blocked by buildings
or planting on the side opposite.
The roads do not continue
straight across, J, K, L and M.

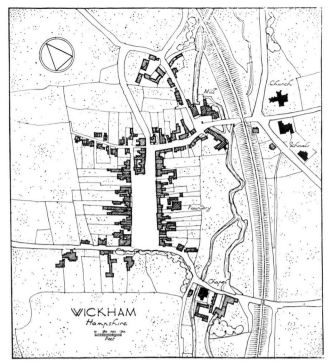

2.150

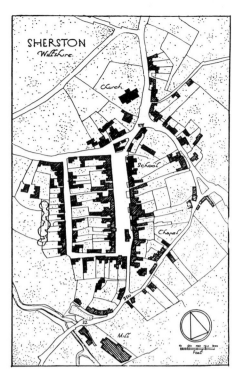

2.153

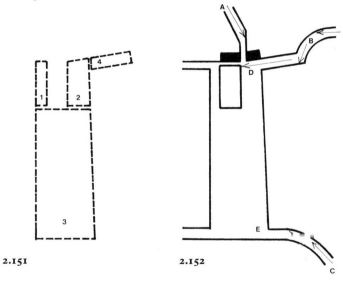

2.151

2.152

2.154

The space enclosed by the village buildings at Wickham, in Hampshire, **2.150**, is a simple rectangle in plan but the existence of an island site within the primary space causes it to be subdivided into three interconnected rectangles, 1, 2, and 3, **2.151**, of different proportions, thus increasing the diversity and interest of the village. The proportion and configuration of the road entrance at the south-east corner ensures that another distinct space, 4, is enclosed. The views in are stopped before reaching the square, A, B and C, **2.152**. And there are no views across the square which are not contained, D and E.

The plan of Sherston in Wiltshire, **2.153** could be viewed as a variation of the type represented by Wickham but with the island site grown to a much greater size in proportion to the overall area. Consequently the main part of the village consists of two parallel spaces, A and B, **2.154**, one wider than the other and comprising the main village square and the other with the proportions of a road but of limited length, having both ends terminated, one by a T- and the other by an L-junction.

Again all views are contained. The road in from the north widens as it reaches the square but a small island site divides the route and blocks the view in, D. The structures of this island site form the north side of space A and the east side of space C. The view in from the south, E, is blocked by the building sitting in the square and the view out by a T-junction.

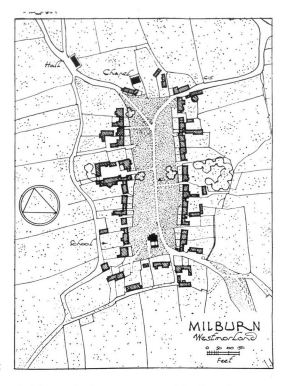

2.155

Milburn in Westmorland, **2.155**, is an example of the type of village which is constructed round an open green rather than a hard surface as in the previous cases. The containment of the views in and the views out have been realized here also and the entrance in from the north west is curtailed so that the person arriving experiences a sudden transition from the narrow approach to the broad green. Equally narrow is the entry to the court on the south side of the square. This court is a secondary outside space situated on the route connecting the centre of the green to the road running outside the village parallel to the south side.

One of the unsatisfactory features of the roads in modern developments is their similarity from one area to another and from one part of the country to another. Of the 10,000 English villages no two are the same. These villages constitute an invaluable store of examples embodying good townscape principles which will handsomely repay study by designers concerned with creating good road form. A number of principles contributing towards the success of the village layouts illustrated can be listed as follows:

1 The layout is such that the village road or square is experienced as a place. It is experienced as having an outside and an inside.

2 The place is a volume contained by the grassed or hard floor and by the sides of the buildings.

3 The thresholds to the place are of special importance. The routes in and out are so arranged that all views are contained. The points of entry are often marked by a narrowing of the road thus creating a portal effect.

4 The layouts are simple in the sense that there is one clearly primary space in each village. Secondary spaces, when they exist, are obviously subordinate.

5 When more than one outside space exists they tend to be dissimilar in form, maximizing contrast and interest.

6 Building lines are often not parallel and not straight. The same is true of road edges. Normally there are no very large differences in building height except in the case of special buildings like the village church.

7 Special buildings are usually placed in a dominant position where they act as a visual focal point and often as the fulcrum on which the total composition of the layout is balanced.

3
FORM

Static and dynamic space

We have seen in our discussion of networks how the environment can be conceived as a system of places connected by routes. This way of experiencing our surroundings is related to the concept of static and dynamic space. A static space is one which, by its form, conveys a sense of rest and completeness, whereas a dynamic space implies movement and change. A static space tends to be circular or square and associated with 'place' and a dynamic space tends to be linear and associated with 'route'. There is no precise correlation; we are talking about tendencies only. Roads, which are routes, also have characteristics of places to a greater or less degree and townscape alignment is concerned with increasing the sense of place.

In the layout of the urban environment the simplest groupings of buildings, where order exists, are the cluster and the row or terrace. These too can be correlated with static and dynamic space provided the reservation stated is maintained and consideration of the static and dynamic characteristics of roads in the townscape context must also involve reference to clusters and terraces, **3.1**.

With some layouts it is a simple matter to decide whether the static or dynamic qualities of a space are dominant: a Georgian square is clearly a static space and an avenue is dynamic. But many spaces do not fall so easily into one or other category yet all must lie somewhere along the spectrum whose theoretical extremes are the purely static and the purely dynamic. This spectrum is represented graphically in **3.2**.

Since good townscape alignment is concerned with creating a sense of place it must also aim to reduce the dynamic and increase the static aspects of the space through which the road travels. We will now look at some ways in which this can be done.

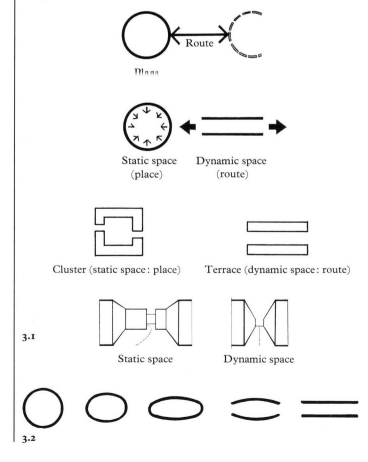

Route

Place

Static space
(place)

Dynamic space
(route)

Cluster (static space: place)

Terrace (dynamic space: route)

Static space

Dynamic space

3.1

3.2

3.3a

b

c

3.4

Considering first a straight row of terraces, the treatment of the elevation and floorscape will play a large part in determining the rate at which the space appears to recede into the distance. If the accent is placed on horizontal lines along the length of the road, the rate of recession will increase; if it is placed on vertical lines on the elevation and transverse lines at the road surface this will result in an apparent slowing of the rate of recession, **3.3a** and **b**.

In the example of **3.3c** the horizontal lines on the elevation are completely dominant, whereas in **3.3d** the dividing walls between the dwellings have been exposed to introduce a vertical element which counterbalances the horizontal emphasis of the windows.

This aspect of design has been considered under the heading Straight Alignments in Chapter 2, Elements of Alignment. It need only be emphasized here that in the townscape context it is important that the configuration of the enclosing sides reinforces the static quality of the space being traversed by the road. This can be achieved in many ways, one of these being to ensure that the openings in the elevation are in groups of different size and alignment, as in **3.4**. Here the different treatments of the adjacent elevations in terms of texture, shade, colour and so on assist in attaining the desired end.

3.5

3.6

The containing sides

In considering the containing planes at the side of the road it must be mentioned that it is not only the vertical and horizontal emphasis which is important in determining whether or not the space appears static and restful or dynamic and restless.

Whatever lines and areas are dominant will influence this issue. The psychological and emotional impact of line and pattern is a vast subject which can only be referred to here, but two simple examples are presented to underline the need for designers to give due consideration to this important topic. The first example shows the elevation of a terrace of Georgian town houses, **3.5**. The well-balanced system of rectangles implies stasis. In the second example, **3.6**, a typical development of modern detached dwellings, the strident, jagged progression produces an impression of disjointed motion.

Although a great deal can be accomplished by the use of line and pattern on a flat elevation these are by no means the only ways by which the sides of the routes can be modified. By engaging the third dimension of depth new possibilities are introduced.

One way of realising these possibilities is to project part of the buildings a short distance into the contained space, producing a kind of bas-relief effect, **3.7**.

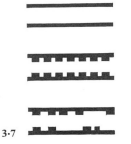

3.7

3.8

3.9

The use of bay windows has been common in this regard. If identical floor plans incorporating bay windows are repeated along a terrace an external regular pattern is set up, **3.8**. This provides a vertical emphasis, but a horizontal element is also introduced by the regular repetition. The result can be a pleasing rhythmic background to street activities and the diversity of small front gardens and street trees. However if the terrace is too long and elements introducing diversity are lacking the effect can be monotonous and overbearing.

This difficulty can be tackled by avoiding the unvaried repetition of identical units. This has been achieved in **3.9** by employing handed floor plans and interspersing flat-fronted buildings with those displaying bay windows. Other methods of providing projections have included the use of balconies and porticos as in **3.10**.

3.10

Taking the articulation of the vertical plane a stage further, the volumes of the dwellings or parts of the dwellings can be made apparent from the road by varying the distance from the road centre line to the fronts of consecutive buildings or parts of buildings. This can be carried out in a regular or irregular manner, **3.11a** and **b**. In example **3.12** a regular stepping forward of the individual dwelling has been introduced part of the way along a terrace resulting in a powerful sculptured effect where otherwise there would have been only a flat surface.

The alternate use of two different sized building units, with one stepped back, exposes to view the volumes of the dwellings, **3.13**. A less formal effect has been produced in **3.14** where the repetition of units is less apparent.

In all these examples the linear aspect of the road is reduced by the creation of small sub-spaces which penetrate transversely into the containing sides. At the same time the horizontal lines on the elevation are broken up and so do not contribute to the effect of parallelism.

3.11a

b

3.12

3.13

3.14

One particularly dominant line of the simple terrace which is interrupted by the process described is that of the eaves, **3.15**. An interesting and varied roofscape, achieved by such means as variations in the roof height and pitch and the use of setbacks and dormer windows, can contribute greatly to the character of a road and thereby to its being perceived as a place.

As can be seen from photograph **3.16** such variations can be highly effective in reinforcing the uniqueness of a scene even when there is little other modulation in the vertical plane. It is important that, as in this example, the variations are not sufficiently great to destroy the unity of the whole and that all the elements remain in harmony.

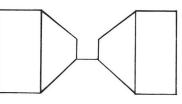

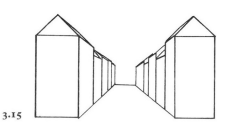

3.15

16

Interconnected spaces

So far we have considered modifications to the linear space of the road by the articulation of the elevations and small discontinuities in the building line. However the structures lining a route can be laid out to deviate substantially from a common building line parallel to the road so that the juxtaposition of forms creates a series of distinct spaces, **3.17**. Illustration **3.18**, which is part of a sketch plan used in the Essex Design Guide, shows the formation, in this way, of a series of spaces along a route.

The construction of individual spaces along a road is often achieved by quite simple means. Slight alterations in the road width between buildings on opposite sides, for example, brought about gradually or abruptly can have a marked effect on the local character of the road, **3.19**. This point is illustrated by photograph **3.20**, in which variations in the width form two distinct volumes, one between the viewer and the zebra crossing and the other between the zebra crossing and the tower. The siting of this tower makes it a powerful focal point for this length of road (see also Chapter 2, Elements of Alignment) and by registering in the mind as a goal and a stopping point it reinforces the image of the length of road as a place rather than a route. The surroundings of the road can be laid out so that quite distinct and large scale static spaces are formed which the road passes through. The roadscape will be greatly enhanced if the route is thoroughly integrated into these spaces so that they are not something which happens outside the road but are an essential part of a person's experience as he journeys along the route. The possibilities are endless, but a few examples will help to underline the attraction of this aspect of road design. This serial linking of spaces can be carried out in a formal or an informal manner, **3.21a** and **b**.

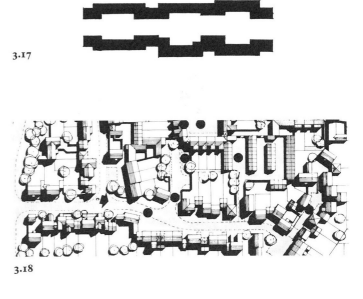

3.17

3.18

3.19

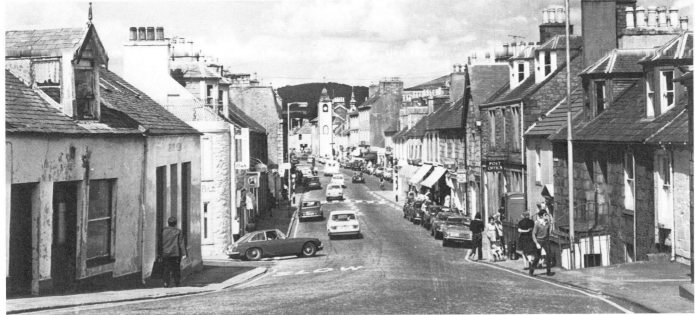

3.20

3.21a

b

The aerial photograph, **3.23**, of Randolph Crescent, Ainslie Place and Moray Place in Edinburgh is a superb example of the formal approach. The layout compels a person moving through the complex to trace out the shape of the contained spaces, **3.22a**. In the unthinkable event of an alternative route being forced through the middle of the layout, **3.22b**, the experience of the spaces would be quite different and much less complete.

By way of contrast to the previous formal example from Edinburgh New Town, the layout of the Royal Mile, **3.24**, in Edinburgh Old Town is much less formal and achieves the incorporation of static spaces by quite different means. Road broadening and narrowing, small changes in direction of the building lines and widening out into a large space (in which is sited the High Kirk of Saint Giles) all contribute to the fine buildings forming a series of unsurpassed outside spaces.

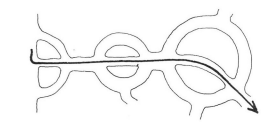

3.22a

b

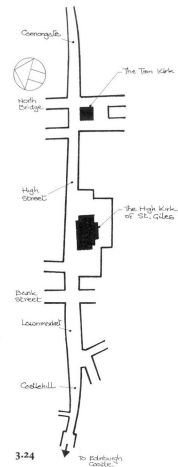

Cannongate

The Tron Kirk

North Bridge

High Street

The High Kirk of St. Giles

Bank Street

Lawnmarket

Castlehill

3.24

To Edinburgh Castle

.23

99

3.25

3.26

Looking west from the
intersection of Bank Street and
Lawnmarket towards the entrance
to Edinburgh Castle the 17th-
century tenements narrow down
the road as the goal is
approached, **3.25**. As the castle
draws nearer, the church on the
left thrusts out into the road, and
from that part onwards the
narrowness is such that the
buildings appear to press in on
either side, heightening the drama
as the threshold to the castle is
reached, **3.26**.

Looking east from Lawnmarket the view is closed by the gradual curve of the left hand side of the road and by St. Giles jutting out into the High Street. That part of the road between the viewer and the church is seen as a single circumscribed volume tending to static space proportions. The sense of containment is assisted by the foreshortening effect of perspective, **3.27**. The four tier system of kerbs provides the pedestrian with a dominant position in relation to the traffic, at the same time facilitating his descending to road level when necessary. Travelling further along the road the 15th-century lantern tower of the church moves into view from behind the flanking buildings on the right-hand side.

The closer the church appears the more it seems to move out into the road, blocking the route and dominating the view, **3.28**. At the same time the large open area in front of the church becomes apparent on the right and invites investigation. At this point the route itself takes on for the moment a secondary rôle. It can be seen to continue on past the church but on a greatly reduced scale. Even so another focal point has already come into view, in the form of a church spire, inviting further exploration.

3.27

3.28

In both the Old Town and the New Town examples from Edinburgh the interconnection of spaces has been made by means of linkage in series. This matter has been discussed in Chapter 1, Networks, but it is raised again here because of the intimate connection with the experience of space as dynamic or static.

3.29a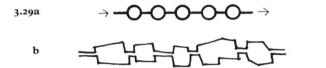

b

A form of connection in series is represented diagrammatically in **3.29a** and **b**, in which a series of static spaces are connected by a series of dynamic spaces. As has been seen the static spaces can have formal or informal plan shapes. Static spaces need not necessarily be joined by linear spaces; they may be joined in series directly like the links in a chain without the use of separate connecting elements, **3.30a** and **b**.

3.30a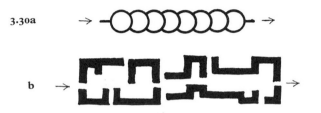

b

Where places are joined in series it is necessary to traverse them all in order to go from one end of the route to another. By connecting in parallel instead of in a series, a choice of routes is introduced and it is no longer necessary to traverse all the static spaces, **3.31a** and **b**.

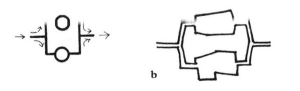

3.31a   b

A combination of serial and parallel linkage results in a limited choice of routes, **3.32a** and **b**.

3.32a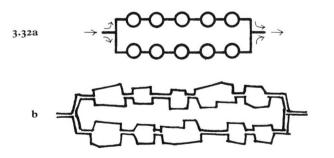

b

The static spaces will convey a greater sense of place if they are not linked in a straight line as in **3.32** but have a more organic arrangement as in **3.33**.

A much greater choice of routes becomes possible with cross connections between elements in the series, **3.34**.

3.33a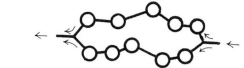

b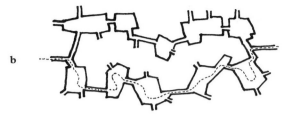

3.34a

b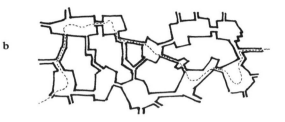

This type of layout can be added to indefinitely and its complexity and multiple choice of route can result in stimulating and varied environments, **3.35**.

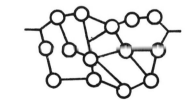

3.35a

b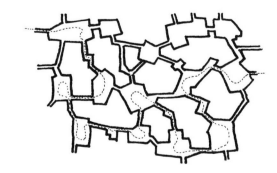

When the static spaces of **3.32** are connected by elements at right angles to the routes shown, a rectangular grid results. These also can be added to indefinitely but the sense of place will be weakened if the routes run straight through. There is a continuous spectrum of organisational possibilities between a rectangular grid and a completely random development.

A plan of the centre of Rome made by Nollis in 1748 provides an example of a layout in which a great variety of spaces are interconnected to form a complex asymmetrical network. The outdoor spaces have been as carefully arranged as those inside the buildings, 3.36 and 3.37. The way the map is drawn showing the spaces inside the buildings as well as those outside testifies to an appreciation of their intimate inter-relationship. A great number of devices has been used to create good townscape, from slight widening and deflections in the roads to dramatic contrasts in scale, shapes and volumes in the outdoor environment.

3.36

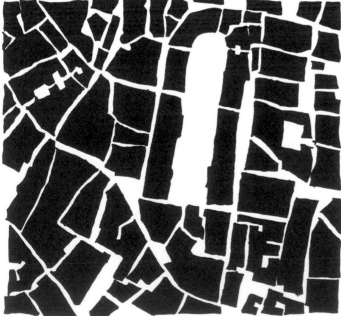

3.37

Relating forms

A length of road consisting of a sequence of spaces, whether static or dynamic, should be composed into an integrated whole. Each space should be related to the adjoining spaces. This relationship may be in terms of contrast, so that interest is heightened by moving from a small space into a large one, from a dynamic space into one which is static, from an area of hard landscape to one containing vegetation. On the other hand the relationships may be established by subtle changes in proportion or variation in the materials which contain the spaces.

This length of road should, in its turn, be experienced as bearing a definite relationship to its surroundings. This relationship may be one which is immediately evident to a person passing along the road, for example a common view, say of a tall building, which can be glimpsed in different aspects from each of the spaces in the sequence; or it may be one which requires experience of a local environment extended over a number of journeys. A person moving for the first time through a labyrinthine area of narrow streets may not be aware that close by is a wide straight avenue. When he has learned of this thoroughfare the knowledge will colour his experience of the narrow streets on his next visit. A few ways in which a length of road can relate to neighbouring areas are indicated below.

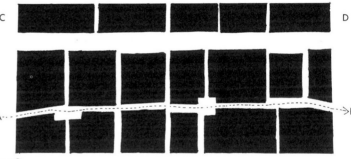

3.38

A narrow road composed of straight lengths but with slight occasional differences in orientation and variations in width is shown in 3.38 and is labelled AB. These ingredients combined with small open spaces along the route, well-proportioned cross sections and interesting containing elements will afford an attractive sheltered road with closed views ahead. The character of this road will be heightened by the existence of a parallel road which is wide and straight and is designed as a thoroughfare or avenue, CD. Periodic transverse roads joining AB and CD will make the traveller aware of the parallel and contrasting alternative way and when moving from the enclosed narrow route with its restricted views into the one which cuts through the urban fabric he will experience a dramatic change in orientation.

In the previous case the two roads were closely related by their linear character. A greater contrast to the straight wide route can be created by an adjoining serpentine one which links in series a number of static spaces, route AB in 3.39.

Alternatively a route linking a sequence of large-scale spaces can be contrasted with an adjacent labyrinthine area, 3.40.

When different aspects of a dominant building or view are glimpsed along a route a sense of continuity is experienced

and the viewer feels well orientated in his surroundings, **3.41**. When this view appears as a goal being approached the experience of arrival is intensified if the object disappears temporarily and then suddenly reappears in close up, **3.42**.

3.39

3.40

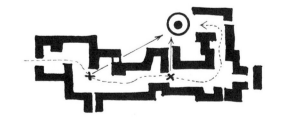

3.41

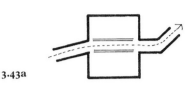

3.42

The recurring prospect need not be that of a single object. It could be for example a series of views of a linear park or of a nearby river. When travelling along Western Road in Brighton a view of the sea occasionally appears along a transverse street. This increases the sense of being sheltered from the elements in the urban surroundings. It is important that such views out are restricted so that the feeling of being sheltered is not lost.

104

Experiencing the static space

The way in which a road passes through a static space will condition the extent to which that space registers with the observer. If it passes through a cluster of dwellings but is flanked by high walls, shrubs or rows of closely spaced trees the cluster will not register and the route will be experienced as dynamic, **3.43a**. Whereas if a more open and varied treatment is given to the edges, **3.43b**, and any immediate containing elements are kept low, then the overall space will be dominant. Two examples of this are given in the section Straight Alignments in Chapter 2, Elements of Alignment, photographs **2.21** and **2.22**. Another way of ensuring that the space is experienced as a whole is to take the route round the perimeter, as described above for Randolph Crescent in Edinburgh, **3.43c**.

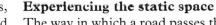

3.43a

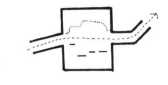

b

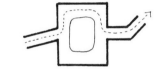

c

Closing the view

The strongest sense of containment is achieved in a space when the views out and those in are closed. A view through a space provides no inducement to pause when the space is reached, **3.44a**, whereas a closed view implies a stopping point and the necessary change in direction induces a reduction in speed. This closure can often be brought about by staggering the entrance and exit, **3.44b**.

A simple but highly effective instance of this occurs at Blanchard, Northumberland, **3.45**. As can be seen, a slight staggering of the two openings in the containing terraces is sufficient to produce the desired effect. Also taking the route through an archway as has been done at one of the entrances is a most effective means of inculcating a sense of arrival in a particular place. This is a device which can often be simply included in modern developments. The use of an archway entrance has already been illustrated in connection with the mews, in Chapter 2.

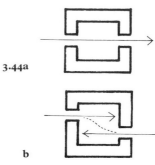

3.44a

b

3.45

105

Other methods of closing views have previously been mentioned, namely T-junctions, curved alignments, **3.46a** and angled alignments, **3.46b**. A striking example of a curved alignment used for this purpose occurs at Finchingfield in Essex, **3.47**. Photograph **3.48** shows how the view out is completely blocked by the road curvature with the result that the structures appear to present a continuous enclosed wall. It is noteworthy how the apparently simple juxtaposition of individual buildings, as seen in plan, produces in perspective what seems an uninterrupted, containing elevation, as illustrated on the right hand side of the photograph. The church has been masterfully sited outside the square but on a dominant position and seems to entice the traveller along the road towards its half-hidden presence.

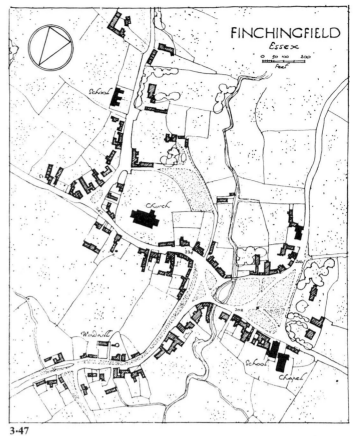
3·47

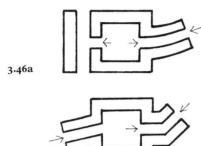

3.46a

b

3·48

The size of outdoor spaces

The space inside a building must have a size and proportion suitable for its intended use and the dimensions must be such that a person feels comfortable in the space when it is being used for its specific purpose. A family of four persons would not feel comfortable dining in a banqueting hall designed for a hundred people. For both these numbers there is an optimum size of room. If the space in either case is too small the people will feel cramped, if it is too large a lack of intimacy and a feeling of being in an unprotected environment will result. Similar considerations pertain to the size of outdoor spaces. In *Exterior Design in Architecture*[6] the Japanese architect Yoshinobu Ashihara formulates what he calls his 'one tenth' theory, on the basis of his own experience. His theory states that 'In the design of exterior space a scale that is about eight to ten times that of interior space is adequate'. As an example he cites the Japanese $4\frac{1}{2}$ mat room. A mat is the Japanese unit of measurement for floor space and is approximately $1\cdot8 \times 0\cdot9m$; so that a $4\frac{1}{2}$ mat room is $2\cdot7m$ square. Using the one tenth rule an equivalent outdoor space would be 21.6 to 27m square. In a space of this size even when furthest apart two people would easily be able to recognize each other and an intimate exterior space will result. A further example given is that of the 80 mat room ($7.2 \times 18m$) which is a spacious room often designed for banqueting halls. In Japan it is traditionally considered that this is the maximum size of room in which people can interact in an informal manner. Applying the one tenth theory again, but multiplying by eight, the equivalent outdoor space becomes $57.6 \times 144m$, which he suggests is possibly the largest size space in which some sense of intimacy is maintained. This view is reinforced by Camillo Sitte. In *The Art of Building Cities* he says that the average size of large plazas in European cities is $58 \times 142m$. Ashihara points out that such a rule should not be slavishly adhered to and can be varied from one fifth to one fifteenth, depending on the requirements of a particular design. What is important is to keep in mind that such relationships exist between interior and exterior space and to design accordingly.

Certain dimensions related to the individual should also be borne in mind by the designer of outdoor spaces. At 1200m a man can just be detected, at 24m he can be recognized and at 14m his face can be clearly seen. At 1–3m he is seen in direct relationship to the other person. The nature of the experience of this relationship varies from discomfort if the proximity is forced upon strangers, to feelings of intimacy between friends.

Outdoor spaces with a dimension of 3m are considered as being very small, 12m appears intimate and up to 24m still maintains a very comfortable human scale.

PROPORTION AND SCALE

Proportion concerns the relationship between different dimensions of a space or object. It is a relationship which is independent of size. The sides of a square have the proportion 1:1 irrespective of the actual size of the square. Scale deals with the relationship between the size of a space or object and some other space or object outside itself. A building can be said to be in scale or out of scale with its surroundings whether they are urban or rural. If the building is in scale it is considered that its size and shape

relate well to the context. The most important scale is the relationship between the size of spaces and forms and the human figure. This relationship is a powerful factor in determining the character of an outdoor space and the way it is experienced by an observer.

The human eye is reputed to have a normal field of vision of about 60° although 45° is about the limit of the range in which any degree of detail can be seen.

When the height to width ratio of a route exceeds 1:1 a canyon effect begins to be experienced. A person looking with a level gaze at the opposite elevation will only take in about a quarter of the height if the height to width ratio is 1:0.25, **3.49**. In the alley of **3.50** this ratio is 1:0.17.

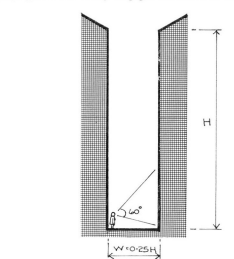

3.49

3.50

When the height to width ratio is 1:0.5, about half the height of the opposite elevation can be easily seen, **3.51**. This ratio is 1:0.6 in **3.52**.

These proportions accord a high sense of containment and of being protected from the elements but they also tend to be overpowering and to produce feelings of claustrophobia if they are sustained. They are therefore best used over limited lengths of road, as a contrast to lengths with more generous proportions and as a tightening of space before the sudden release of moving into a more open area.

At 1:1 there is a good balance between building height and the space between. A comfortable space is created. A 45° cone of vision extends from the bottom to the top of the opposite elevation from the far side, **3.53**, but it is still difficult to see the elevation over its full height. A high sense of enclosure is still felt. Both the road in **3.54** and the mews court in **3.55** have a height to width ratio of 1:1.2. The entrance to the mews in **3.56** also has a ratio of 1:1.2.

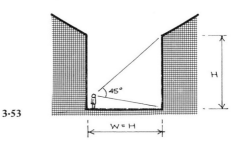

3.53

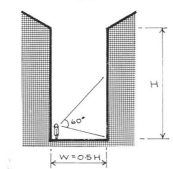

3.51

3.54

3.52

108

3.55

At 1:2 it is easy to see the opposite side of a road over its full height, **3.57**. The entire elevation and its details are perceived clearly and the elevation will almost fill the field of vision. The street shown in **3.58** has a ratio of 1:2.3.

At 1:3 a building seen at the opposite side of the space will still dominate the field of vision but if it is part of a group it will be seen as such. The cone of vision encompassing the elevation is reduced to 18°, **3.59**, the sense of spatial enclosure is quite low, and the elevation begins to lose detail. A row of buildings will begin to function predominantly as an edge to the space, providing a sense of place rather then a vertical containing element.

If the buildings are not conceived with a view to structuring the space between them then the degree of spatial definition will be very low even when the width is not very large in relation to the height. The height to width ratio in **3.60** is approximately 1:3.5.

3.56

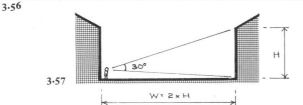

3.57

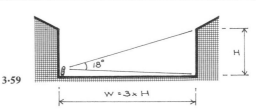

3.59

3.58

3.60

At 1:4 a dwelling would appear as part of a general scene and a row of dwellings would be seen as a band in the cone of vision which would contain a large proportion of foreground and sky, **3.61**.

In the layout of buildings the ratios 1:1, 1:2 and 1:3 are often the most satisfactory. Ashihara says 'When D/H becomes equal to 1, then we feel a balance between building height and the space between buildings. In the actual laying out of buildings D/H=1, 2, and 3 are most frequently applied, but when we exceed D/H=4, the mutual interaction begins to dissipate . . .'.

The chaos which can result from ignoring the importance of well-proportioned spaces contained by visually satisfying forms is indicated in photograph **3.62**. Here the height to width ratio is 1:10. A sight like this is not uncommon in modern suburban residential developments. The various height to width ratios are summarized in the diagram below, **3.63**.

3.61

3.62

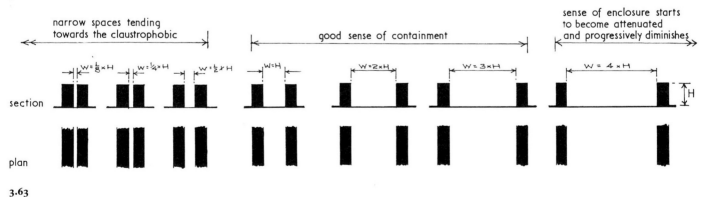

3.63

Form and speed

The control of rate and direction of movement are primary considerations in the design of outdoor spaces. The lines, areas and shapes of townscape elements can imply a speeding up, slowing down, constant speed, constant rate of change of speed and so on. Some of the relevant characteristics of form have already been mentioned but it is convenient to summarize them here. This is done with the aid of the simple diagrams below, **3.64**. In this section we are considering only the impact of different configurations on the behaviour of the observer. (Obviously all designs must give due consideration to safety and traffic circulation. Thus although for example converging sides to the road will tend to induce a reduction in vehicle speed such a device is only likely to be applicable in terms of the enclosing elements and not in relation to the running surface except perhaps for very slow speeds.)

When the sides of a route converge the subjective impression of speed is heightened and a slowing down is induced. This is the case whether it is the road surface which narrows or the containing sides of the space which come closer together. A gradual narrowing will tend to produce a gradual slowing down. Sometimes a road will contract abruptly, but if this can be seen from a distance a gradual slowing will result in this case also. If the narrowing is temporary, for example due to a structure coming to the edge of the road when those on either side are generally set back, there will tend to be a slowing down followed by a build-up of speed as the obstacle is passed.

Conversely as the road and the space containing it widen the driver experiences a feeling of release and will increase speed. For a particular width of road the further the structures are from the road edge the higher is likely to be the driving speed and vice versa.

A straight line implies a state of tension and is conducive to high speed; a dotted line implies a constant slow speed. This could be translated into physical terms, for example in a short cul-de-sac, by having alternate areas of the road surfaced in different materials, colours or textures.

Parallel lines in the direction of travel, whether on the vertical or horizontal planes, will, as we have described, encourage high speed.

Vertical lines on the vertical planes and transverse lines on the horizontal planes will encourage slow speeds.

Patterns which set up small-scale rhythms in the vertical or horizontal planes encourage slow speeds.

Whereas those patterns which produce large scale rhythms encourage high speeds.

A line of this type implies varied speeds and disjointed movement. It can be appropriate when it is necessary to keep the speed low.

A series of small-radius curves of equal amplitude and length imply a constant slow rhythmic movement as in the meandering bends of a slow-moving stream. The smaller the radius and the greater the amplitude, the slower the motion.

A constant decrease in radii implies a constant decrease in speed and vice versa.

Long reverse curves are compatible with high controlled speed at a constant rate.

An inconsistent line of this type appears weak, lacking in energy and purpose. Speed on such a line is likely to vary.

A slope upwards is associated with slow, controlled speed.

A slope downwards denotes high speed which is difficult to control.

Motion in a clockwise direction appears to imply greater speed than motion in an anti-clockwise direction.

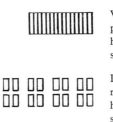

4
ELEMENTS OF TOWNSCAPE

The attributes of good townscape operate at many levels, from the superficial to the profound. They are all concerned with the feelings and emotions evoked in the individual. It may be a mild feeling of pleasure produced by the texture of a cobbled surface or, at the other extreme, it could be a deep spiritual feeling inspired by the symbolism of emerging from a dark narrow road into the light filled square in front of a great church.

At whatever level the surroundings act on the viewer it is always necessary for the designer to instil visual interest and variety into the outside spaces.

Whatever the purpose of a particular design it is necessary for the designer of outside spaces to have a vocabulary of elements which he can use to evoke a specific response. Such elements should always be employed in a meaningful manner to attain a particular objective. A few which have been used in the past and which relate particularly to roads are briefly referred to here with photographs of one or two of the many ways in which they have been integrated in the townscape scene. How they are utilized in future developments will depend on the needs and life style of the road users and the skill and knowledge of the designer who should add to those elements listed by his studies of specific localities.

The concept of townscape is developed in *Townscape*[7] by Gordon Cullen and it is on that work that this section is based.

The lines and planes of the sketches indicate the perimeter of the volumes through which the road passes. The edge of the road surface would not necessarily be parallel to the building lines of the sketch plans. And, of course, the bare outlines of the spatial volumes and enclosing planes indicated in the sketches must be clothed in such a manner as to produce an integrated whole with its own character and style.

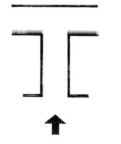 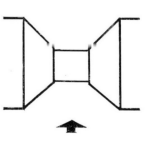

4.1

T-junctions
The T-junction is a classic townscape method of closing views and thus helping to create a sense of place, **4.1** and **4.2**. From a townscape point of view a T-junction differs in important respects from a Y-junction. Obviously the closure is more abrupt with the T-junction and although alternative routes are presented to the user, these are in exactly opposite directions. With the T-junction there may be only a small angle between one route and the other. From the nature of the T-junction the alternative routes as seen from the intersection tend to be part of the same space, and thus they tend to have the same proportions and character. This is less likely to be the case with Y-Junctions. Since the two routes are not continuous with each other their character and proportions may be, and often are, quite different. Also it is often possible to see the entrance to the routes of the Y-junction from some way off, so that the choice may be presented gradually and not suddenly as with the T-junction, **4.3**.

4.2

Y-junctions

Y-junctions present a clear offer of a choice of routes. **4.3**. The character of the alternatives can often be seen to be different, thus stimulating interest in other parts of the neighbourhood and inviting exploration. As with all the other townscape elements mentioned, Y-junctions are used in a great variety of scales and styles of urban fabric, **4.4** and **4.5**. The absence of offensive street furniture in Arundel, **4.5**, enables the buildings to impose their stamp on the character of the place unhindered by extraneous distractions. This is not the case in the London street scene, **4.4**.

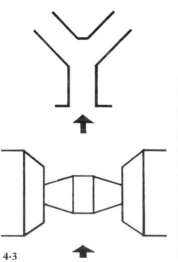

4.3

4.5

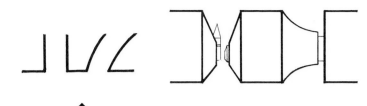

Multiple views

Being able to see two places at once gives the viewer a pleasant feeling of omniscience. It also enables spaces of different form and character to be compared, not in series as is usually the case, but simultaneously, **4.6**.

In the Cambridge street scene, **4.7**, the linear road is in complete contrast to the court which can be glimpsed through the doorway to the left of the picture. The dynamic and static spaces are seen together, exposing to view how one generates the other. The buildings which envelop and form the courtyard with one elevation also contain the linear space with the other elevation. Thus by seeing two views at once a person is better able to comprehend the townscape.

4.7

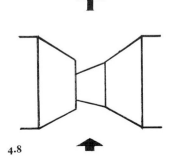

4.8

Angles

A variety of impressions can be conveyed by a change in angle of the road alignment. When the angle is small, a long view remains even when the vista is eventually closed by the containing buildings, **4.9**. In this example the circular corner tower in the centre of the photograph appears as a fulcrum on which the locus of the road centre line rotates.

4.9

Curves
The properties of curved roads have been considered in Chapter 2, Elements of Alignment. This is another townscape element which has had many and varied applications. As well as closing the view a regular curve appears to draw the viewer into and around the enclosed volume, **4.10** and **4.11**.

In this example the repetition of vertical elements—window and door openings, masonry brackets and railing uprights—modulate and add a powerful rhythm to the façade, whose strongly emphasized horizontal lines sweep the eye round the curve.

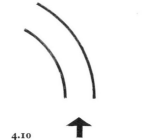
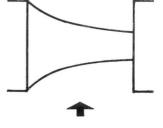

4.10

4.11

4.12

The pivot
The road in **4.12** appears to pivot about the towering tenement block on the left. This is more than just the appearance of an alignment avoiding an obstacle: the road seems to be an integral part of the building about which it rotates.

Deviation

Even a small deviation in the line of the route divides it into distinct places. At the same time it reduces the look of urgency of a straight alignment and pushes structures into the line of sight which would otherwise be much less prominent, **4.13**.

Deflection

A structure whose axis is at an angle to the primary direction of a route can appear to deflect the user into a new direction, **4.14**. This device can be employed for a range of changes in orientation from a few degrees to a right angle. A slight change in direction is indicated by the white building in the distance in **4.15**, whereas a much more abrupt and immediate change is enforced by the deflecting structure in **4.16**.

4.13

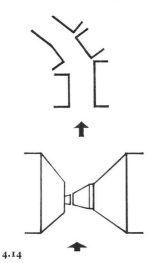

4.14

4.15

4.16

116

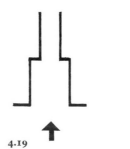

4.17

4.18

Level Change

Movement into a higher position or descending to a lower one is always an occasion for an interesting townscape arrangement.

Descending has many connotations, one of which is a movement from an exposed position down into a more enclosed and protected environment. The high walls containing the lane of **4.20** and its serpentine path intensify the sense of descent and at the same time impart a feeling of expectancy as the lowest level is approached.

Another way of using change of level to good effect is to present in one view the choice between the higher route and the lower one, **4.17** and **4.18**.

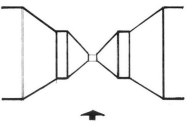

4.19

4.20

Fluctuation
Fluctuation in the width of the route is of major importance in the differentiation of space and has been discussed above in Chapter 2, Elements of Alignment, and Chapter 3, Form.

Narrowing
In the first illustration, **4.20**, the road surface extends to the vertical planes containing the space, so that stepping in the building line has meant narrowing the road surface. A common solution is to maintain the same road width despite the alteration in the building line, **4.21**.

Road narrowing need not always be carried out by alterations in the building line, **4.22**. Street markets provide traditional examples of dual use areas in which vehicles and pedestrians successfully share the same surface. It is tacitly agreed that pedestrians have priority and that vehicles must circulate at walking pace. It is worth noting that this state of affairs often exists without the need for extensive signposting and without the legal enforcement of a 3mph speed limit.

4.22

Funnelling

Instead of narrowing abruptly the reduced width of the containing space may be brought about gradually. If the narrowing takes place at a more or less constant rate this will tend to distort the perspective. As a result the space, when viewed from the wide end towards the narrower end, will appear to be longer than it is and when viewed in the opposite direction it will appear to be shorter, **4.23**.

An example of funnelling is illustrated in **4.24**.

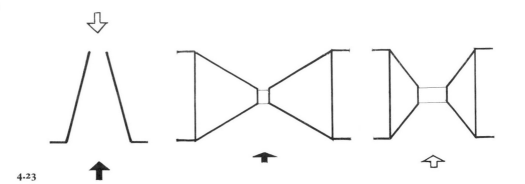

4.23

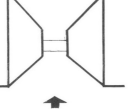

4.24

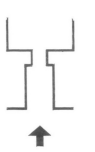

4.25

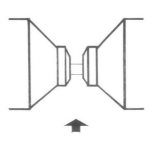

4.26

Widening

Movement from a narrow space into a more open one engenders a pleasant sense of release, **4.25**. Emerging from an urban street into a square or from a lane onto a village green or simply from a narrow length of road into a wider one can all be enjoyable experiences. These urban experiences have countless parallels in nature; a path through woodland becomes a clearing or a narrow valley emerges onto a plane.

Constriction

The experience of contrast between the two spaces can be intensified if the approach is narrowed just before the more spacious area is entered. It is as though the constriction builds up pressure, making the subsequent sense of release more complete, **4.26**.

Wings
In this configuration structures
aligned more or less at right
angles to the axis of the road are
analogous to the wings of a stage,
4.27 and **4.28**. It is as though the
road were a platform onto which
the various actors emerge from
the side to enact the drama of the
street. Endless layout possibilities
exist in the formation of wings.
The device can be used to bring

about a slight narrowing in the
route, creating a visual obstacle
which must be negotiated, **4.29a**.
Or the building can thrust out
across the road leaving only a
minimum gap, **4.29b**. The road
can be narrowed down gradually
by a series of wings, **4.29c**. An
angled building can appear as one
of the wings at the same time as
forming an interesting urban
space, **4.29d**.

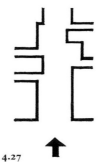
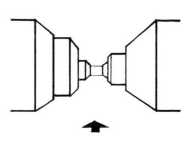

4.27

4.28

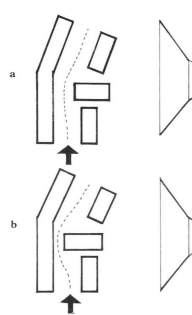

a

b

4.29

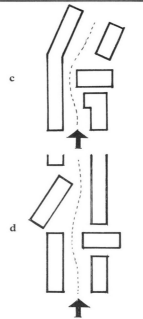

c

d

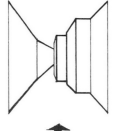
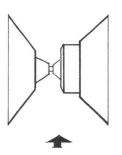

120

The chasm

There is something exciting about squeezing through a cleft in a rock mass or travelling along a deep chasm. The excitement has an element of fear that the rock may move but it also includes a feeling of exhilaration that the fear is being overcome by the knowledge that this will not happen. There is also a sense of awe at the scale of the threatening mass of material. Similar sensations can be awakened in an urban setting when buildings are sited close together with narrow routes between them, **4.31**.

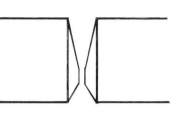

4.30

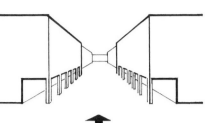

4.32

4.31

4.33

The Colonnade

This time-honoured addition to the street scene has many advantages. It creates a protected environment for the pedestrian parallel to the road. He is physically protected from the weather but also the rows of columns impart the feeling of being protected from the traffic on the roadway. The vehicles become less obtrusive, **4.33**. Visually the vertical lines of the columns act as a brake to the converging lines of a straight road and the enclosed space appears more static than would otherwise be the case. It is also possible that rows of columns have pleasant associations of woodland scenes for the viewer, of which he may or may not be conscious.

The overhang

A sense of excitement is generated by the sight of an overhang of rock on a cliff face. Buildings which overhang a route can produce a similar sensation. Even when the overhang is from only a small part of the building it can be a significant townscape element, since if well sited it can be a prominent element in the view for some time for the approaching motorist or pedestrian, **4.34** and **4.35**. Overhangs can appear threatening or protective depending on their mass, form or proportions and even the most temporary projection can add to the enjoyment of the streetscape, **4.36**.

4·35

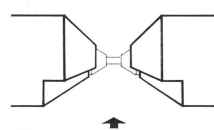

4·34

4·36

The arch

Whether it is located at the entrance to a mews or a city, the arch is always a powerful symbol of entrance to a place. The act of 'going under' the arch as well as 'going through' is experienced and not just 'going between' as would be the case if the route simply entered a road flanked by buildings, **4.37**.

4·37

4·39

38

The bridge

The bridge as a townscape element can be used in many different ways. Going under a bridge can denote entering or leaving a place in a similar manner to going through an arch.

In **4.38** it is used to dramatic effect since it is necessary to descend under the low soffit of the bridge deck in order to emerge beside the massive overshadowing building.

At a smaller scale, passing under the room bridging over the service vehicle access and footway in **4.39** emphasizes the exclusiveness of the space being entered.

Passing over a bridge is always a positive townscape experience for a pedestrian. But the motorist is not always aware when he is on a bridge and the interest of a route is increased if it can be arranged so that the driver can see the structure from the approach road.

123

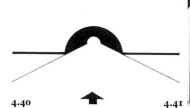

4.40

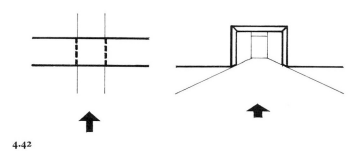

4.41

The maw

A dark tunnel-like entrance is both forbidding and enticing. It stimulates our curiosity at the same time as it makes us feel slightly apprehensive. This element of townscape appears in many guises; the entrance to an underpass, a road in a tunnel, through a building, under a viaduct and so on, **4.40**.

The black hole which tunnels through the edge of St. James' Palace, **4.41**, intrigues and attracts despite the outcome of taking that route being perfectly evident.

Going through

To pass through an opening in a structure is to instinctively heighten one's awareness. The opening formalizes the passage from one space into another. But it also restricts the view of the space which is about to be entered. The movement is from the known into the unknown, **4.42**.

When the opening has the noble proportions of that in **4.43** the sense of imminent change is intensified.

4.42

4.43

Closure
This occurs when the containing elements bound the space yet the sense of progression along the route is not lost, **4.44**, and the form of the space evolves as progress through it is made.

Enclosure
Whereas in the case of closure the dynamic element is still quite apparent, with enclosure it is almost totally subdued. The viewer finds himself in a static space, the form of which remains unaltered as it is traversed, **4.45**.

4·44

4·45

Going into

A portal to an outside space performs a similar function to a doorway into a room. A person goes through a doorway expecting to step into a contained space. Portals to outside areas can be designed to engender the same expectancy. The portal should imply containment and a certain degree of privacy. The actual degree of privacy can vary over a wide range. The gateway into a walled town implies privacy but one which is shared by all the inhabitants of the town. An archway into a courtyard may herald a private zone which is shared by only one or two households. The scale of the portal should be commensurate with that of the space to be entered or the importance of the entrance.

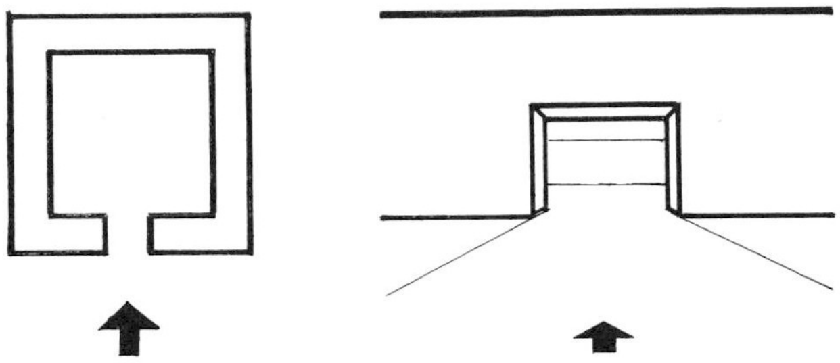

4.46

The entrance shown in **4.47** leads the driver or pedestrian into a landscaped area overlooked and shared by a large number of the residents of the Lillington Street Development in Westminster.

4.47

Dead end
In townscape terms this does not
simply mean that there is no exit.
This fact should be absolutely
apparent. Such a dead end can be
built to different scales, but the
message of 'No through road'
should be as unequivocal as in
4.48.

4.48

Hinting
In the design of a road for fast-
moving traffic it is important that
the driver is never faced with an
ambiguous situation. With a
flowing alignment the route must
be absolutely clear; one curve
must lead on to another in a
manner which is foreseen. It is
even preferable that the
subsequent stages in the
alignment should appear to follow
an inevitable logic. This is not
necessarily the case with the
townscape alignment. Situations
arise in which the next move is
only hinted at by what can be
seen from a distance. This can
arouse curiosity and a desire to
explore. Motorists will approach
such an area with caution.
The sketch **4.49** indicates only
one of the many layout
configurations which result in the
person entering a space being
given only a hint of the way in
which the next space in the
progression is to be entered.
Photograph **4.50** illustrates such a
situation. The opening on the
right becomes more obvious as
the space is penetrated.

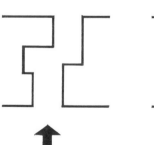

4.49

4.50

4.51

Enticing

This townscape feature occurs when the view along the route includes an element which is, at the same time, attractive and not directly attainable. In **4.51** the viewer is enticed forward to explore the route and try to find a way to the church, of which only the tower is visible.

Isolation

A dramatic effect can be achieved when the road passes a structure apparently isolated and viewed against the sky. In **4.52** the impression is heightened for the pedestrians by the route being aligned with the gable wall before turning into the space between the facing terrace blocks.

The Argyll Hotel in **4.53**, although apparently isolated, is the first building in a continuous row which forms one of the containing sides of the street seen receding into the distance. By approaching the road in this way the full impact of the contrast between the vast open spaces across the Firth of Clyde which are visible to the right before passing the hotel and the enclosed, sheltered space of the street, can be fully appreciated.

4.52

4.53

Framing

The same structures which contain the route can be used to frame a building or group of buildings, **4.54**. This device can be used to enhance the appearance of the structures at the same time as it makes the road more interesting. It is applicable to any urban scale; an avenue can be used to frame a palace or a narrow road to frame a small church as in **4.55**.

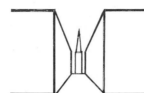

4.54

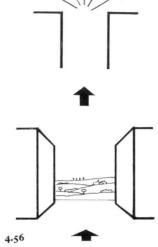

Vistas

Where the urban fabric meets the countryside an abrupt transition is usually more pleasing in appearance than a gradual one, **4.56**. The alignment of roads can often be chosen to take advantage of this circumstance by presenting to the traveller a dramatic view of the countryside being approached. Such a view may be flanked by the buildings not yet left behind, **4.57**. Use can be made of changes in ground level to expose a vista to best advantage, **4.58**. Of course, vistas and panoramic views of the urban fabric itself can be presented by careful arrangement of routes.

4.56

4.55

4.57

4.58

4.59

4.61

Incident

The view of a road can be enlivened by some designed 'happening' along its length which, although not substantially altering the form of the enclosed space, is sufficiently marked to create a temporary focal point. This incident can be, for example, a structure which stands out from its surroundings by virtue of its unusual façade, as illustrated by the Tudor frontage of the building at the bend of the road in **4.59**.

Punctuation

The division of the linear space into separate, serial, entities can be achieved by structures or planting which, because of their form or position, clearly indicate the termination of one space and the beginning of another. Alternatively they may be used to register the occurrence of a transition zone.

In **4.60** the tower, as well as providing a focal point for the main village street, marks the change from the wide central zone to the narrower width prior to leaving the village. An even more emphatic punctuation is achieved by the building facing down the street in **4.61**. If this building was lined up with the others the viewer would perceive the space as continuing as it curved round the corner and out of sight. This structure, by its orientation, indicates a 'full stop' at the top of the hill before the next space is entered.

Landmarks
Travelling through an urban area
can be a disorientating
experience. If a particular
structure, by virtue of its
prominent siting, can be seen
from different locations this will
greatly assist the traveller in
maintaining his bearings, **4.62**.
Landmarks can fulfil this function
in different ways. It may be that a
particular structure appears, is
lost to view and then reappears as
a particular route is traversed. In
other cases a prominent building
will be seen on different journeys
from different routes in an urban
area, thus assisting the observer
to gradually build up a mental
picture of the area.
The church of St. Margarets,
4.63, magnificently sited on a
hillside over-looking Runcorn
New Town, can be seen from
many different routes and assists
in the orientation of pedestrian
and motorist alike.

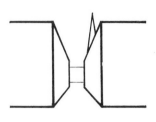

4.62

4.63

Elements of townscape—summary

Junctions

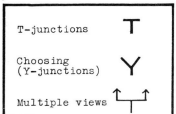

T-junctions

Choosing
(Y-junctions)

Multiple views

Line

Angles

Curves

The pivot

Deviation

Deflection

Level change

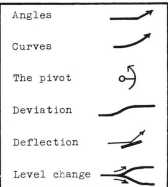

Width

Fluctuation

Narrowing

Funnelling

Widening

Constriction

Wings

Overhead

The chasm

The collonade

The overhang

The arch

The bridge

The maw

Going through

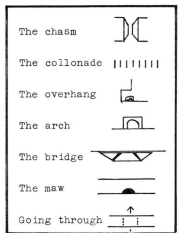

Features

Hinting

Enticing

Isolation

Framing

Vistas

Incident

Punctuation

Landmarks

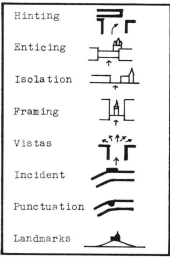

Containment

Closure

Enclosure

Going into

Dead end

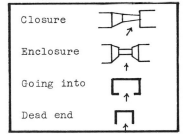

PART TWO

THE FLOWING ALIGNMENT

To the eye of the moving observer, the highway slab and its shoulder form an unwinding ribbon of parallel lines, swinging and changing into various horizontal and inclined planes, standing out in stark white or black against the soft, warm colours of the landscape. As it turns and changes direction, as it rises and falls over hills and valleys, as it diverges to accept a stream or pulls together to enter a city, the paved ribbon assumes qualities of an abstract composition in space, which gains in richness because it is not only passively seen but actively traversed by the driver, who experiences visual as well as kinesthetic sensations of tilting, turning, dropping and climbing.

Tunnard and Pushkaref, Man-made America

Our inter-urban highways, designed for fast moving traffic, flow across the landscape, their shape dictated by the physical laws of vehicle movement, the capabilities of the machines and their drivers, and the need to fit the line to the land form.

In this part attention is drawn to those alignment forms which fulfil these requirements in the most aesthetically pleasing way. A well designed road will possess an internal harmony—the alignment will be pleasing to the eye of the driver as it unfolds before him; and it will possess an external harmony—it will be arranged in such a way that those using the road will be presented with interesting and coherent views out to the landscape being traversed. A good flowing alignment on high-speed roads will mean safer, less stressful driving conditions as well as being visually satisfying.

The flowing alignment can be applicable to the design of distributor as well as trunk roads and certain aspects of this type of line are relevant to attaining the best appearance in the design of some access roads. The avoidance of short straights between circular curves, for example, can lead to an improved appearance on minor roads and the use of transition curves can result in a better-looking line although they may not be necessary to satisfy the dynamic requirements.

Two chapters are presented in this part: Internal Harmony and The External View. The first lists the linear elements of a flowing alignment and goes on to consider how these elements should be joined to produce the best driving conditions and most visually pleasing line. The laws governing vehicle movement, an important factor in determining the form of the alignment, are examined in Appendix A, The Dynamics of Alignment. The second chapter is concerned with the external harmony of the alignment and information relevant to this topic is also presented in Part III, Land Form and Planting.

5
INTERNAL HARMONY

The line of a road is conditioned by the way in which vehicles move and by the ways in which the driver experiences this vehicle movement. When travelling at high speed the driver must be able to see a long distance ahead and the road must flow in a predictable manner. To avoid skidding and for the comfort of those in the vehicle, centrifugal force must be applied gradually and be kept within a predetermined limit. The fulfilment of these requirements is totally compatible with the design of an aesthetically pleasing road; in fact it leads the designer towards that end. However the application of relevant formulae will not automatically produce the required result. A thorough understanding of the dynamic characteristics of the vehicle (see Appendix 1) and a detailed appreciation of the driver's view of the road alignment are necessary on the part of those involved in the development of the design. This last aspect of design, the driver's view of the road, will now be considered. (The geometry of a particular design should comply with the recommendations of *Roads in Urban Areas*[8] or *The Layout of Roads in Rural Areas*,[9] whichever is applicable; both these documents were prepared by the Department of the Environment and embody government policy on the design of roads other than residential access roads.)

Elements of the flowing alignment

For the sake of clarity all the possible elements are listed. However when we consider their integration into a continuous alignment it will be seen that a three-dimensional approach should be taken to the design and that often the best results are likely to be obtained by a series of three-dimensional curves.

Any line can be considered as a sequence of elements. The primary distinction which can be made is that between straight and curved elements. This distinction can then be refined in the following way.

Straight lengths of road may be elements which are level or on a gradient, **5.1**.

For curved elements there are a number of possibilities.

Horizontal circular curves are of major importance on flat ground, being compatible with change of direction at constant rate.

Transition curves are horizontal curves designed to obtain a gradual change in direction and centrifugal force, **5.1d**. These are used to effect a gradual transition between a straight and circular curve, **5.2a**, or between two circular curves of substantially different radii. If the circular curves are in the same direction one transition curve is required, **5.2b**, but if they are in different directions two are used consecutively, **5.2c**. It is also sometimes possible to dispense with a circular curve and simply to use two transitions, **5.2d**. However this usage is not satisfactory for other than minor, low-speed roads owing to the need for the driver to continually alter the direction of the vehicle wheels as he progresses round the curves. It can also result in the appearance of a kink at the point of minimum radius. Spiral curves are used as transitions and commonly the 'Euler' curve is the one adopted. Its mathematical name is the clothoid. The degree of curvature at any point on this spiral varies directly with the distance along the curve.

Vertical curves are those designed to effect vertical changes in direction. They can be used in series, **5.3**, or in

conjunction with level lengths of road and with gradients, in other words with tangential lengths.

In the latter case the relation between the vertical curve and the tangential length of road may take any one of the four forms indicated in **5.4a** to **d**.

Curves resulting in a convex surface are called summit curves. Those forming a concave surface are called valley curves. Vertical curves can be circular, simple parabolic or cubic parabolic. The simple parabola is generally used. In the design of highway profiles the simple parabola closely approximates to the circular curve.

The three-dimensional curve which results when a straight gradient, ab, is placed round a cylinder is called a simple helix, **5.5**. The resulting curve is circular in plan but takes the form of a gradient in elevation.

The variable pitch helix is a three-dimensional curve resulting when a vertical curve, cd, is placed round a cylinder, **5.6**. The resulting line is circular in plan and takes the form of a vertical curve in elevation.

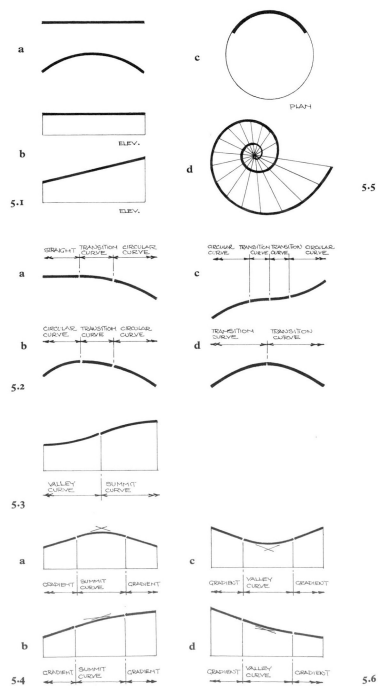

5.1

5.2

5.3

5.4

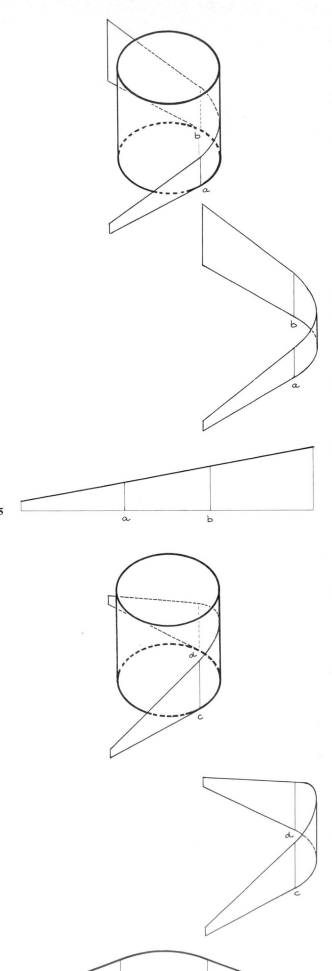

5.5

5.6

135

Flowing alignment and the driver's eye view

The choice and arrangement of the linear elements are crucial factors in ensuring that the road will look right in its surroundings and will be pleasing to the driver of the vehicle, as well as fulfilling its function of providing a safe route.

In the early days of modern road design the line was worked out initially on plan. The vertical alignment was then determined, taking due account of the need to minimize the amount of earthworks and the haulage distance of cut and fill materials. Little attention was paid to coordinating the horizontal and vertical aspects of the design. As vehicle speeds and the scale of the roadworks increased it became apparent that to achieve the best results certain precepts should be observed regarding the sequence of elements and the way in which the plan and elevation of the line should be coordinated. Finally it was realized that ideally a major road should be conceived in three dimensions. It should comprise either a series of three-dimensional curves or, where appropriate, curves of this type used in association with two-dimensional elements. Of course, on perfectly flat terrain the alignment will be two dimensional, but this is quite rare in Great Britain. And so three-dimensional elements, the simple helix and the variable pitch helix, became added to the list of two-dimensional elements which had previously been used. The type of line which results from this evolved approach to design has been called a flowing alignment.

The aim of flowing alignment is to combine the various components in a manner which results in the road being experienced by the road user as a free-flowing, harmonious form without any visual discontinuities. Such a design results in better integration of the road into the landscape than was previously achieved in modern highway layouts and helps to make the road a construction which is visually pleasing from the viewpoint both of its users and those outside its boundaries.

It has been pointed out in the Introduction that the flowing alignment is the type appropriate for fast, major routes in the landscape. Other circumstances call for other alignment types. Nevertheless this approach to road design provides vital insights to the designers of all types of road.

The designer must decide in each case whether and to what extent the principle of flowing alignment is applicable. For a district distributor it may be desirable to use this approach as the dominant theme, **5.7**, whereas in the design of a particular access road it may be a question of using a single helical curve to fit a particular land form, **5.8**.

Although it is essential to the flowing alignment approach that the road design is considered from the outset as a three-dimensional problem, nevertheless in the ensuing discussion it will be convenient to start by considering the horizontal and vertical aspects separately, and then go on to deal with the way in which they should be combined.

The principles of flowing alignment are closely linked with the way in which the driver sees the road line and in particular the shapes of the road edges. This is a matter which has been studied by a number of writers, among whom are J. F. Springer and K. E. Huizinga. In *The road-picture as a touchstone for the three-dimensional design of roads*[10] they describe a method of drawing the perspective view of the road as seen by the driver and of studying the visual effects of different types of configuration.

The diagrams presented in this chapter are not scale drawings but sketches to illustrate the various points being made.

5.7

5.8

THE STRAIGHT

The perspective view of a straight length of road is basically triangular in shape. The triangle is symmetrical if the road is viewed along the centre line, otherwise it is asymmetrical. If the edges of the road are parallel in plan they will appear to converge at a point on the horizon when seen along their length, **5.9**. Across flat terrain a straight alignment can convey a powerful sense of purpose, **5.10**.

THE HORIZONTAL CURVE

These curves are predominantly arcs of circles. If a horizontal circular curve lies entirely in front of the observer it will be seen as an ellipse. If only part of the curve is in front of the observer then it will appear as a hyperbola. The road user is not likely to be aware of the overall shape of the curve until he is near its beginning or within it. Consequently a hyperbola is normally the curve seen, **5.11**.

The perspective view of a transition curve is part of a curve of the third degree. It generally only becomes important when seen from a short distance. From further away its effect on the overall appearance of the road line is negligible.

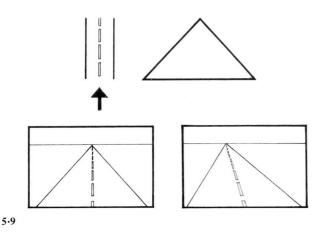

5.9

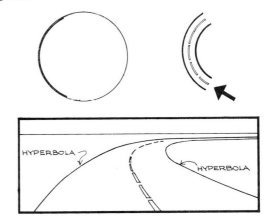

5.11

5.10

THE VERTICAL CURVE

As has been mentioned, it is normal to adopt the simple parabola, **5.12**, for vertical curves. Coordinates of this curve are easily calculated since the equation is $y = \text{constant} \times x^2$.

When driving within a vertical curve the edge lines will appear as hyperbolae to the motorist, **5.13**. The equation of a hyperbola is $y = \text{constant} \times \frac{1}{x}$.

THE COMPOSITE CURVE

This curve bends horizontally as well as vertically. Again the edge lines appear as hyperbolae to the observer. When the vertical curvature dominates, the outside edge will seem concave, **5.14**, and when the horizontal curve dominates it will show up as convex, **5.15**.

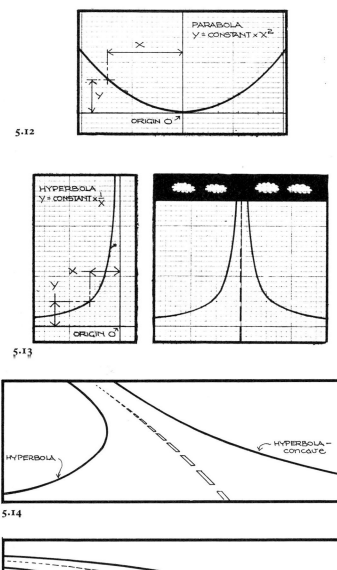

5.12

5.13

5.14

5.15

Combining elements: the horizontal plane

The elements of the horizontal alignment are straights, circular curves and transition curves. In minor roads it is often possible to dispense with the use of transition curves and to connect straights and circular curves or circular curves of different radii directly without the use of an intermediate curve. But for major, high-speed roads they can only be dispensed with when the radius of the circular curve is very large. (For example in *The Layout of Roads in Rural Areas*[11] the minimum radius for curves without transitions is given as 1500m for dual 7.3m carriageways.) For their normal usage, between tangents and circular curves, transition curves are not referred to in the remaining part of this chapter for the sake of simplicity of presentation although in practice they would be provided as recommended in the relevant Department of the Environment documents, referred to above.

It is advisable to avoid small changes in direction in a flowing alignment. These are likely to appear unsatisfactory in the view from the vehicle. Also small transverse displacements can present a confusing prospect for the driver. In **5.16** the line has been displaced by the width of the road by the incorporation of an arrangement of reverse curves.

When two straights are connected, the use of a short horizontal curve is likely to cause the appearance of a kink, **5.17**. In such cases the impression can be improved by employing a larger radius but an improvement only results provided the views being compared are taken at the same distance from the vertex of the curve **5.18**. If the viewpoint is at the same distance from the start of the curve then an improvement is not achieved, in fact the kink will appear to be rather more pronounced.

Even with a large radius curve it is not possible to avoid the illusion of a sharp change in direction if the approach straights are sufficiently long, **5.19**. The best results are likely to be achieved with the flowing alignment when straight lengths can be dispensed with. This of course is not always possible or even desirable. For example in roads which are not dual carriageways, the sight lines on stretches of road where overtaking is permitted must be based on the passing sight distance and not the stopping sight distance. Straight lengths may then be required to achieve these sighting distances. Also it should be borne in mind that such effects will not necessarily be significant in the total view for any particular case. Each design should be considered in its landscape context. This is true of many aspects of internal harmony although the greater the design speed the less the external features modify the internal views, since vegetation and buildings are further back from the road edge, the carriageways are wider, sight lines longer and the roadworks generally are constructed to a larger scale.

Abrupt changes in direction can be unsatisfactory on access roads as well as on highways. In **5.20** the straights have been joined without the use of a horizontal curve. The appearance is quite different when a horizontal curve is added, **5.21**.

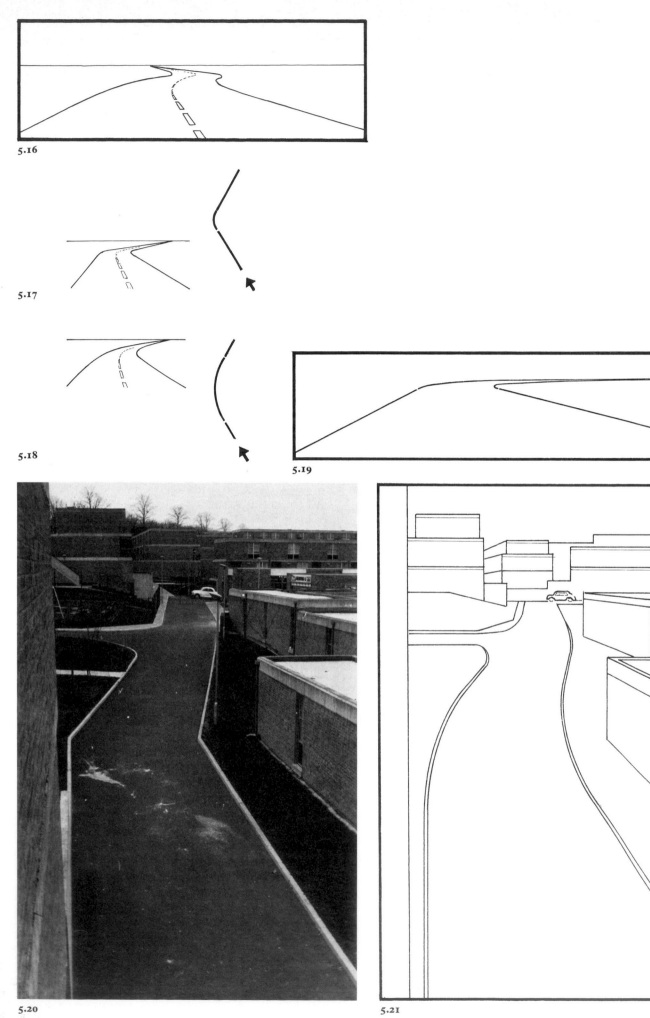

5.16

5.17

5.18

5.19

5.20

5.21

The designer should be aware of the way in which the view of the line changes as the route is traversed. From a distance the change of direction between two straight lengths of road appears to be much greater than when viewed in plan or from a position close to the change in direction. The way in which the apparent abruptness of the direction change alters as the position of change is approached is illustrated in **5.22**. No horizontal curves have been used in these diagrams in order to emphasize the point.

A similar phenomenon is observed when a horizontal curve is approached from a straight, **5.23**.

Short straight lengths of road should not be interposed between horizontal curves of opposite sense since the appearance of a kink is likely to result, **5.24**.

A possible solution is the use of a pair of transition curves, **5.25**.

When designing for slow speeds or in the case of very large radii it may be feasible to join the two curves directly as in **5.26**. This should be done with care since here also an impression of lack of flow may result.

Similarly in the case of two subsequent curves in the same direction the use of an intermediate short straight, **5.27**, is likely to produce an unsatisfactory visual effect. Here there may be the possibility of replacing the two curves and the straight with one circular curve, **5.28**. Another possibility may be to interpose one transition curve between the two radii, **5.29**.

A series of reverse curves is likely to produce a flowing alignment which is pleasing to the eye and comfortable for the driver. This type of line is ideal for integrating a route into an undulating landscape. The fine example of **5.30** illustrates its large scale use. Here the excellent siting results in a road which enhances the beauty of the landscape. An even greater degree of integration would have been achieved had the tree line been carried, with new planting, across the fence and at least part way up the slopes of the embankments and down those of the cuttings. This would have had the added advantage of concealing the visually divisive fence line.

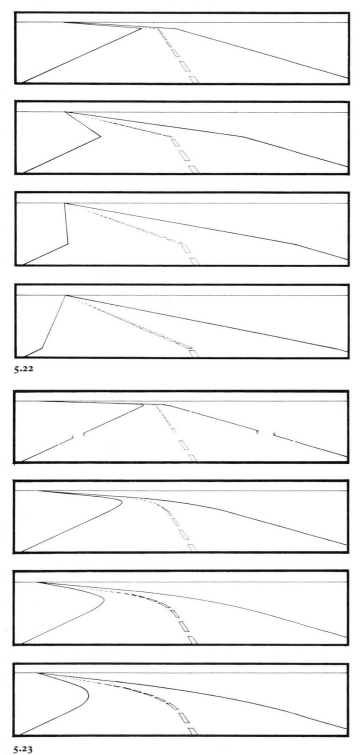

5.22

5.23

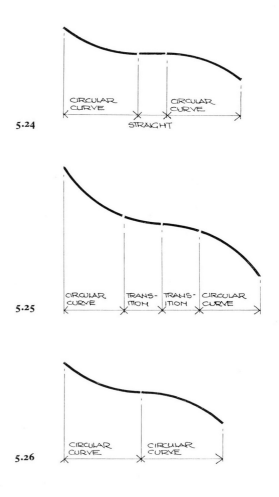

5.24

5.25

5.26

140

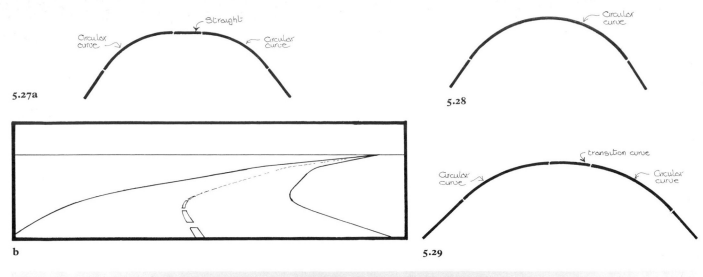

5.27a

Circular curve

Straight

Circular curve

b

5.28

Circular curve

5.29

transition curve

Circular curve

Circular curve

5.30

5.31

5.32

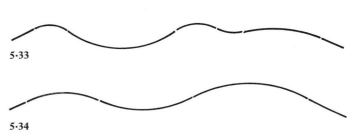

5.33

5.34

Although it was evolved during the development of major highway design, the use of flowing alignment in the form of a series of reverse curves is applicable to other types of road. The local distributor in **5.31** is a case in point and in a small scale a series of curves has been used for the access road of **5.32**.

Of course, in the design of all categories of road, the curves should be used in a meaningful manner and their purpose should be apparent to the observer. The best results are obtained when the curves are well balanced in relation to each other. The aim should be to have sufficient variation in the radii, lengths and types of curve to sustain interest on the part of the road user but at the same time to avoid too large contrasts which would destroy the flow of the line; for example **5.34** as opposed to **5.33**.

The American Association of State Highway Officials (AASHO), in their publication *A Policy on Geometric Design of Rural Highways*,[12] states that 'Consistent alignment always should be sought. Sharp curves should not be introduced at the ends of long tangents. Sudden changes from areas of easy curvature to areas of sharp curvature should be avoided. Where sharp curvature must be introduced it should be approached, where possible, by successively sharper curves from the generally easy curvature'. In the German standards for the design of roads outside built-up areas this desirability of a balanced relationship between successive curves is again recognised. They state that this objective can be attained by keeping successive radii within the ranges shown in the diagram reproduced here, **5.35**. This diagram was developed on the basis of an optical analysis of existing roads and accident analyses.

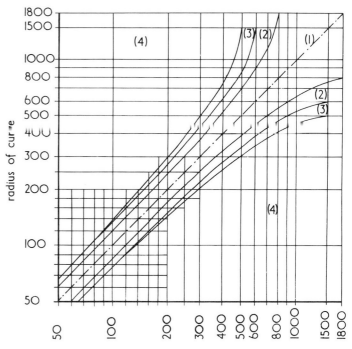

5.35 *Permissable combination of radii (Source: Richtlinien für die Anlage von Landstrasen-Linienführung (RAL-L))*

(1) Very good range
(2) Good range
(3) Usable range
(4) Range to be avoided

142

Combining elements: vertical plane

The concave or valley curve plays an important part in achieving internal harmony in the alignment, especially since it can often be viewed along its whole length at one time. This is not normally possible in the case of summit curves but for this reason particular attention must be paid to ensuring that visual continuity is maintained with this last mentioned type. This leads to the avoidance of short summit curves even though they may satisfy visibility requirements.

As is the case with horizontal curves and straights, when a valley curve is used to join two gradients, or a gradient and a level length, the radius must be sufficiently large for the appearance of a kink to be avoided, **5.36a** and **b**. Even large radii will give the appearance of an abrupt change in direction if the viewpoint is sufficiently far from the curve, but this is unimportant since at great distances it will not be found disturbing. In fact it has been noted that drivers do not tend to become aware of an approaching valley curve until they are about 500m from the start.

Tangents, especially short ones, between two valley curves can result in an awkward-looking line, **5.37a** and **b**.

It has been pointed out that a vertical curve is seen as a hyperbola. Whether or not the junction of a tangent and a vertical curve presents the appearance of a kink depends on the curvature of the sharpest bend of the hyperbola and its location in relation to the end of the tangent. It is desirable that the hyperbola does not start at the position of its smallest radius. In critical cases it is advisable to examine perspective drawings of the line. An indication of the effect of small and large radius vertical curves on the motorist's view are shown in **5.38** and **5.39** respectively.

A level length or gradient containing a short low summit curve can cause a visual discontinuity since the distant length of road, diminished in size by perspective, can be seen over the crest, **5.40**.

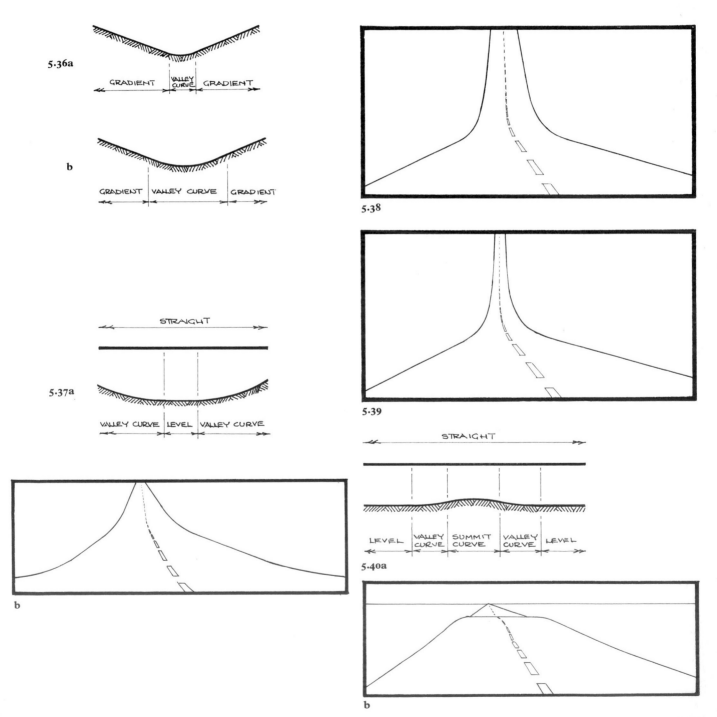

5.36a

5.38

5.39

5.37a

5.40a

143

Similarly the use of a reverse curve in the longitudinal section, causing a small change in level, can result in a visual discontinuity due to the road surface disappearing from view and then reappearing. An example of the effect when a reverse vertical curve is used in conjunction with level straight lengths is shown in 5.41.

A view of this type can occur with a double reverse curve, 5.43. In the case illustrated the line can be improved by increasing the length of the valley curve and decreasing those of the summit curves, 5.44.

When a terrace is created by a sequence of summit and valley curves, whether or not there are tangents between the curves, it is likely to result in an unsatisfactory view if two summits can be seen at the same time. An example is shown in 5.45.

The lower the terrace is placed and the shorter its length the more disturbing it is likely to appear, since it can be viewed from a shorter distance.

All terraces tend to appear unsatisfactory when seen from the top. As with the horizontal alignment, the ideal solution for the vertical alignment is a series of well modulated vertical curves proportioned so that they avoid the problems discussed. Such a solution can, of course, only be used when the land form and other controlling factors make it possible.

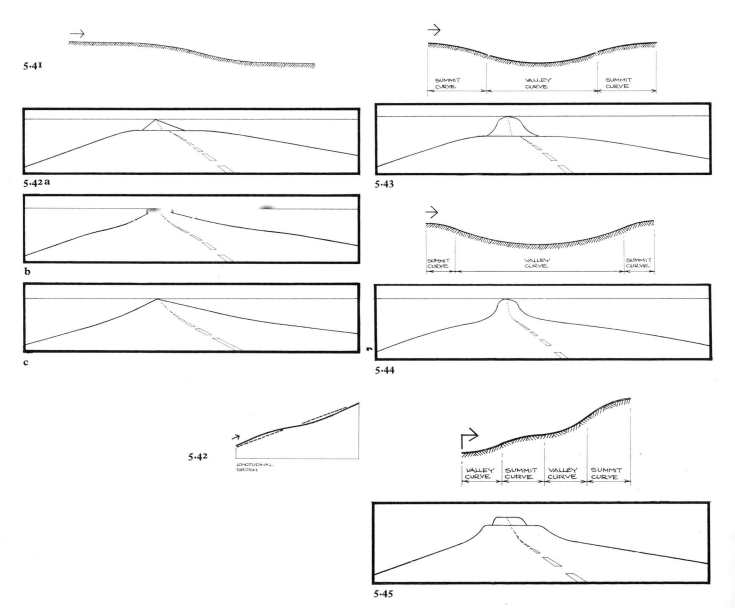

5.41

5.42 a

b

c

5.43

5.44

5.42

5.45

Unifying the horizontal and vertical alignments

So far we have considered the horizontal alignment and the vertical alignment as two separate entities. In fact to obtain a satisfactory line it is most important to integrate the vertical and horizontal aspects of the line. Better still, the designer should learn to think of the road as a three-dimensional unit. This is what it is: one object in the landscape. Vertical alignment and horizontal alignment are intellectual constructions invented to simplify the design and detailing process. Roads are drawn on a drawing board which is flat. This necessitates the projection of the three-dimensional proposed alignment onto a flat plane. A two-dimensional projection of the road profile is drawn together with a plan. It is then possible to fully detail the three-dimensional object, the road, on a flat surface. But the designer must guard against this method of detailing conditioning his way of thinking about the alignment.

We have pointed out that ideally a flowing alignment is made up of a series of well-balanced curves and that where feasible the designer should consider the alignment curves in terms of the simple helix and the variable pitch helix, illustrated in **5.5** and **5.6**. (The three-dimensional curves can be drawn in the traditional manner of employing a plan and longitudinal section. Also the setting out can be done using the standard methods.) Nevertheless two-dimensional curves and tangents are still invariably also required and even with three-dimensional curves it is necessary to examine the relationship between the plan and longitudinal section of the proposed road with a view to obtaining the best flowing alignment.

The elements of the horizontal and vertical alignment should be 'in-phase' wherever possible. In other words the corresponding elements in the horizontal and vertical planes should start at approximately the same points and end at approximately the same points, **5.46**.

If the out-of-phase amounts are small this is not likely to be significant. In fact it is probably advantageous to have some overlap. This is considered by some authorities to contribute to the integration of the two aspects of the line. When an overlap is used it should normally be small in comparison with the length of the element. Yet there are exceptions to this: the plan and profile combination of the type shown, **5.47**, will probably produce awkward-looking perspectives. In this arrangement the horizontal curve ends at the same point as the vertical curve begins.

Whereas circumstances may prevent using longer, coincident curves, by employing longer curves which overlap, as in **5.48**, it may be possible to achieve significant improvement. However the best results would be obtained if coordinated curves of longer radius could be used, **5.49**. The following combinations of horizontal and vertical alignment are some additional examples of those which are likely to result in an awkward appearance.

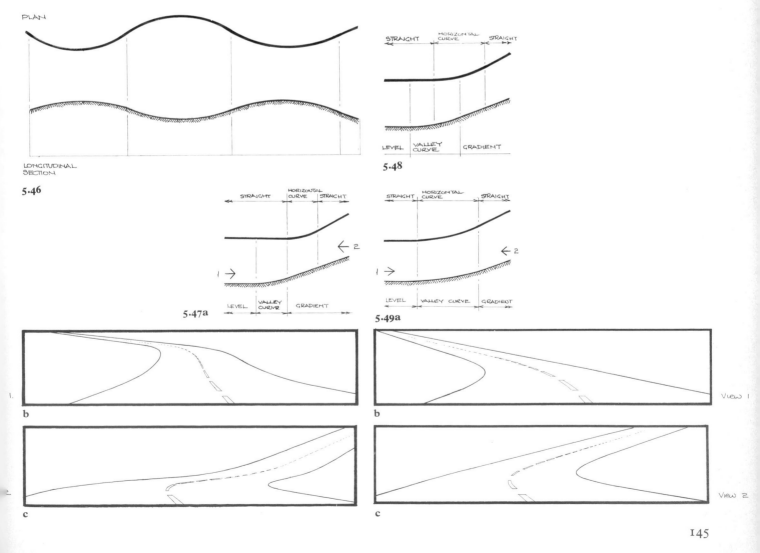

PLAN

STRAIGHT · HORIZONTAL CURVE · STRAIGHT

LEVEL · VALLEY CURVE · GRADIENT

5.48

LONGITUDINAL SECTION

5.46

STRAIGHT · HORIZONTAL CURVE · STRAIGHT

←2

1→

LEVEL · VALLEY CURVE · GRADIENT

5.47a

STRAIGHT · HORIZONTAL CURVE · STRAIGHT

←2

1→

LEVEL · VALLEY CURVE · GRADIENT

5.49a

b

c

b

c

VIEW 1

VIEW 2

Figures **5.50a** and **b**: a short valley curve within a horizontal curve. This is similar to the case of a short valley curve occurring along a straight, but the impression of discontinuity will probably be even more pronounced.

Figures **5.51a** and **b**: low summit curve within a horizontal curve. Here too the discontinuous appearance is liable to be even more pronounced than with the low summit curve on a straight, already mentioned.

Figures **5.52a** and **b**: a short vertical curve connecting gradients in a long horizontal curve. This arrangement is liable to result in the illusion of a pronounced kink in the alignment. Small changes in direction between tangents are undesirable in the vertical plane as they are in the horizontal plane.

Figure **5.53**: a horizontal curve following a straight and starting on a valley curve which follows a gradient. This combination tends to give the horizontal curve the appearance of a sharp bend.

Figures **5.54a** and **b**: valley curves joined by a level length or gradient and occurring along a straight followed by a horizontal curve. Valley curves joined by a tangent are undesirable in themselves but when combined with a horizontal curve in this way they can produce the results shown.

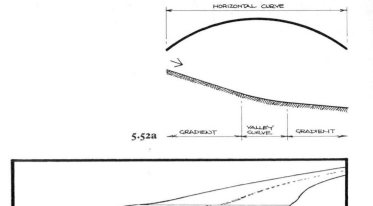

5.52a

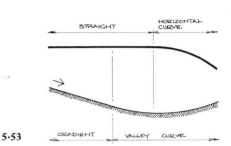

b

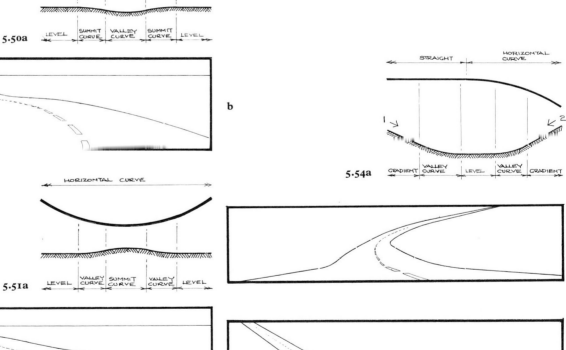

5.50a

b

5.51a

b

5·53

5·54a

b

b

Figures **5.55a** and **b**: a summit curve followed by a valley curve occurring along a straight followed by a horizontal curve. A disjointed effect is liable to result when the beginning of a vertical curve is hidden from the driver by an intervening summit while the continuation of the curve is visible in the distance beyond.

Figures **5.56a** and **b**: a tangent length between a vertical curve and a compound curve. Wherever possible such a tangent length, ab, should not be used. Instead the vertical alignment should be so arranged that the curves can be joined directly.

Figures **5.57a** and **b**: a short horizontal curve within a long valley curve. This combination can result in the appearance of a kink.

Figure **5.58**: a short horizontal curve occurring on a short summit curve. This can be dangerous since the driver is unable to see the continuation of the curved horizontal alignment. An even more unsatisfactory case would be if the horizontal curve started immediately over the summit.

Figure **5.59**: a reverse horizontal curve with the change in curvature situated at the top of a sharp summit curve. This, also, is a dangerous arrangement since the driver is not able to anticipate the change in curvature.

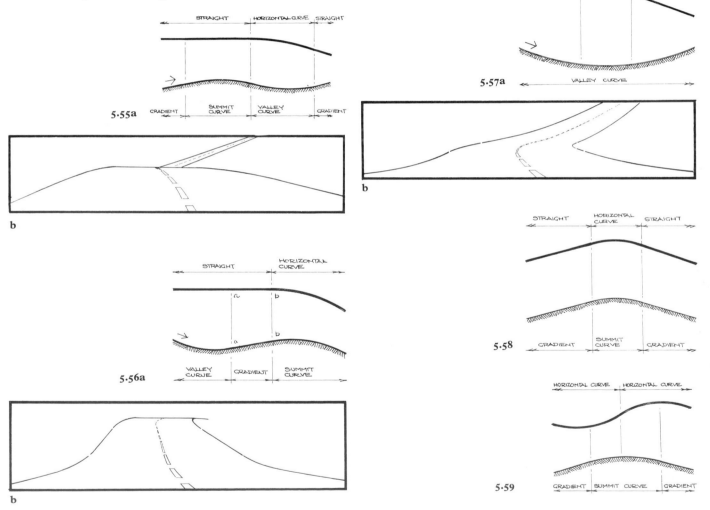

Horizontal alignment—summary

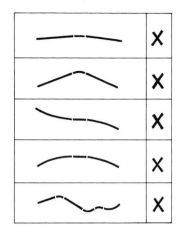

To be avoided:

1 Small change of direction

2 Short horizontal curve between two straights

3 Short straights between horizontal curves of opposite sense

4 Short straight between horizontal curves of the same sense

5 Out of balance alignment

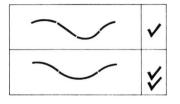

To be attained:

1 Well-balanced alignment

2 Use of curves rather than straights where feasible

Vertical alignment—summary

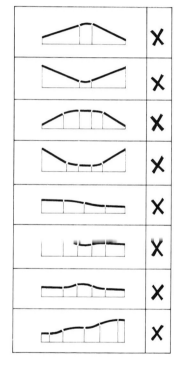

To be avoided:

1 Short summit curves between gradients

2 Short valley curves between gradients

3 Short tangent between summit curves

4 Short tangent between valley curves

5 Reverse vertical curve causing small change in level, on a level length or gradient

6 A level length or gradient containing a low valley curve

7 A level length or gradient containing a low summit curve

8 Terracing on which two summits can be seen at one time

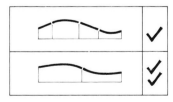

To be attained:

1 Well-balanced alignment

2 Use of curves rather than straights or gradients where feasible

Unifying the horizontal and vertical alignments

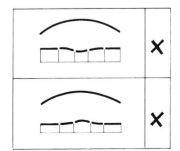

To be avoided:

1 Horizontal curve containing a low valley curve within its length

2 Horizontal curve containing a low summit curve within its length

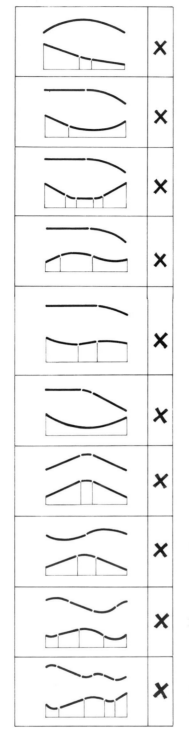

3 Short vertical curve between gradients in a horizontal curve

4 Horizontal curve following a straight and starting on a valley curve which follows a gradient

5 Valley curves joined by a level length or gradient and occurring along a straight followed by a horizontal curve

6 Summit curve followed by a valley curve occurring along a straight followed by a horizontal curve

7 A tangent length between a vertical curve and a compound curve

8 Short horizontal curve within a long valley curve

9 Short horizontal curve occurring on a short summit curve

10 Reverse horizontal curve with the change in curvature situated at the top of a sharp summit curve

11 Out of phase alignment

12 Badly balanced arrangement

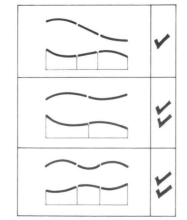

To be attained:

1 Horizontal and vertical curves in phase (the visual continuity can often be improved by having the horizontal elements slightly leading the vertical ones)

2 Where possible use three-dimensional curves and avoid the use of straights

3 Use a well-balanced three-dimensional alignment

6
THE EXTERNAL VIEW

The way in which a person sees his surroundings when moving along a flowing alignment at, perhaps, 70 mph is quite different from the way the environment is seen in a townscape alignment traversed at slow speed. In the former case the lines of sight are very much longer and distant views are the rule rather than the exception. A hedge parallel to the road edge will appear at a distance as a thin, almost static, dark line. Nearby it will rush past, a blurred green surface. In the townscape form the distant view would probably not be obtained.

Looking out from the inside of a vehicle the occupant is able to enjoy a particular type of visual experience which relates to the nature of the vehicle and its path. The viewer is encapsulated in a machine which substantially insulates him from encountering the outside world except through his sense of sight. His senses of hearing and smell are only marginally used and that of touch comes into play only indirectly through the controls of the machine. He does not directly experience the weather. He can feel the motion of the vehicle through his body during periods of change in velocity; otherwise all his attention is concentrated on what he sees. Add to this the elements of speed and a flowing line and conditions are created which demand a special approach to design with respect to the external views. The primary experience is that of motion and space in a constantly changing scene.

Whereas with the townscape alignment it is always clear that the observer is moving through a static environment, in the case of the flowing alignment the illusion is easily gained that the elements of landscape are moving in relation to the driver and to each other. This aspect of a design requires special study if full advantage is to be taken of the opportunities latent in new projects. Some aspects of this and other topics relating to the views out from the road are outlined below. This subject has been dealt with in some detail elsewhere notably in *The View from the Road* by Appleyard, Lynch and Myer.[13] This monograph was written in relation to the design of limited access highways in the United States, with particular reference to the urban context. Nevertheless, as the authors point out, much of the material is applicable to other types of road.

Perception

The mode of perception in human beings conditions what is seen when the observer is moving at high speeds. This subject was studied by Hamilton and Thurstone who presented their findings in a paper entitled *Human Limitations in Automobile Driving*.[14] They put forward five propositions derived from the principles of vision:

1 *As speed increases concentration increases.* The faster the speed the more imperative the need to concentrate attention on the road ahead. Thus, although a road should provide variety and interest, it should be free from extraneous detail and distractions which would divert the driver's attention from the view of the road. It follows from this that views should be developed along the line of the road.

2 *As speed increases the point of concentration recedes.* The driver examines the road ahead at a sufficient distance to enable him to take evasive action if necessary. The distance travelled during the reaction time and braking time increases with speed. At 25 mph the driver will concentrate

on an area 180m ahead, at 45mph this distance increases to 370m and at 60mph it will go up to 610m. Since it is necessary at the higher speeds to focus attention at a greater distance the need for greater concentration (see proposition 1) is partly due to the limitations of human eyesight with the attendant difficulty of making out warning signs in the critical zone.

3 *As speed increases peripheral vision diminishes.* As speed becomes greater the angle subtended by the field of vision reduces. At 25mph the horizontal angle determining the field of vision is about 100°, at 45mph it is 65° and at 60mph it is reduced to less than 40°. This factor also implies that for the driver the main visual events should occur along the axis of the road. If the view encompassed by the narrow field of vision is unchanging, as on a long straight road, this can help to undermine the driver's sense of speed and may induce an inappropriate sense of repose.

4 *As speed increases foreground detail begins to fade.* The faster the objects in the foreground are passed, the more they become blurred and the further ahead the driver must look to obtain a clear image. At 40mph foreground closer than 25m cannot be clearly seen and at 60mph this distance increases to 33m. Beyond about 430m the eye cannot make out detail because it becomes too small, so that at 60mph vision is only entirely satisfactory over a range from 33m to 430m in front of the vehicle and over a width determined by a 40° angle of vision.

This reinforces the view that details in the surroundings are not meaningful to the driver on high-speed roads. He will register the large-scale elements: the land forms, large areas of vegetation, the corridor formed through the landscape by the road itself. All elements of the roadway should be designed in the light of this attribute of perception at speed. Signs should be large, prominent and viewed along the line of the road. Overbridges should have simple, unified forms; intricate details or fine surface textures will not be seen.

5 *As speed increases space perception becomes impaired.* This is particularly true with modern well-sprung vehicles moving on the smooth surfaces of present-day roads. There is no increasing vibration to act as a gauge of speed. Also with large radii, even at high speeds the centrifugal force may be imperceptible. In such circumstances and on straights the main method of gauging speed is by change in apparent size of objects in view and by changes in their position relative to the driver. Since, with increased speed, the area of concentration becomes further in front of the vehicle it is more difficult to perceive these changes and thus estimate the speed. In addition when a driver has been travelling at high speed for a considerable time, and the vehicles in his vicinity have been travelling at similar speeds, he loses his sense of fast motion and his awareness of the time and distance needed in which to stop and take evasive action.

Another factor increasing the difficulty of judging speed on modern highways is the large proportion of the view which is occupied by sky and road surface, the sides being relatively inconspicuous. In the case of a tree lined two-way road the sky and road surface would occupy about 18% of the total view through the windscreen whereas on a six lane highway these elements would take up 60% of the total view. If the reduced field of view resulting from a speed of 60mph were taken into account then the sky and road surface would occupy 83% of the total view in the case of the six lane road.

The sense of speed obtained on a two lane road by the apparent movement of the elements at the side of the route is greatly diminished on a six lane highway, not only because of the small proportion of the view which is taken up by the sides but also because the containing sides are further from the edge of the road and therefore appear to be more static than they would if they were close to the edge.

Views

In arranging views the needs of the driver should be considered paramount but those of the passengers also require to be taken into account. It has been pointed out that the driver's vision should be directed along the road line by, where possible, arranging the major views in this direction, **6.1**.

6.1

At the same time secondary, transverse views will sustain the interest of the other occupants of the vehicle, although front seat passengers tend to focus on views in the direction of the line of motion. (In the research described in *The View from the Road*[15] it was found that there is a strong tendency for those on the front seat to have their attention captured by objects which are straight ahead; other incidents noticed by those in front tended to be obliquely forward rather than to the side.)

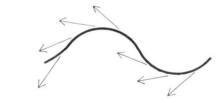

6.2

Changing views avoid boredom and these are best achieved by the scanning line of sight natural to the serpentine route, **6.2**. The eye tends to be attracted to the views on the outside of the curves. The result is the alternating of attention from one side to the other, a circumstance which the designer can use to good advantage by focusing on the best views and drawing attention away from those which are less interesting. The line can be chosen so that attention is attracted to a number of views in series, each one being left behind before the next draws the eye, **6.3**.

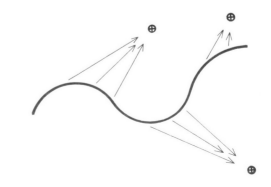

6.3

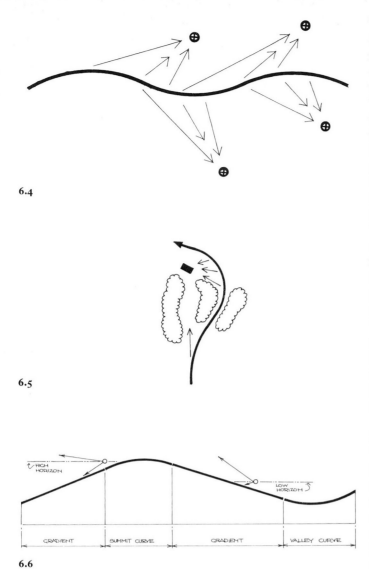

6.4

6.5

6.6

These views may be experienced by the driver as a series of goals. A particular view forward of the vehicle will engage the driver's attention; that view will be approached and passed and then another will become the temporary goal. This succession helps to make the journey eventful and conveys the sense of moving forward to the final destination.

Alternatively an overlap may occur so that one view or goal comes into the line of sight before the previous one has disappeared, **6.4**.

A further possibility is to have a series of short term goals which are established and achieved whilst a more primary long term one is continuously or intermittently visible. When a temporary goal is the object of a driver's attention he is likely to feel a sense of satisfaction if the route finally passes close-by this goal and a sense of anti-climax if it does not.

For a particular view or object there may be an optimum viewing distance which can be utilised in a design. However the road will have increased interest if it is modulated to give close-up and long views in a pleasing sequence. These may be prospects of different locations and objects but there are also cases of attractive distant views being equally attractive in close-up but for quite different reasons, for example due to alterations in scale, field of view, or the new forms which become visible, **6.5**.

The frequency with which successive views appear and disappear, the way in which they are transformed by the motion of the vehicle, the rate at which this transformation occurs, the sequence of near and far views are all factors which can influence the rhythm and tempo of the road. These attributes of rhythm and tempo can greatly influence the character of a route and deserve special consideration in the preparation of a design. They are affected by the internal geometry of the alignment as well as the progression and type of views out. For example, small radius curves, frequent changes in view and large contrasts will increase the tempo whereas slow speeds, large radius curves and wide unchanging vistas will reduce it.

The gradients and vertical curves, also, can be used to display views, especially vistas, to best advantage. When mounting a slope the horizon is low and the field of vision is restricted. The opposite is true when descending a gradient, **6.6**. Here the horizon is high and maximum opportunity is available for the display of panoramic views. The contrast between these two conditions provides many of the most exciting visual sequences on our roads.

In some instances attention can be directed towards the sky as the top of an incline is approached, providing another type of contrast with the vista which comes into view after the summit is reached.

Landmarks

These perform the same function as in the townscape alignment. they assist the driver to orientate himself in his surroundings and enable him to gauge his progress along the route. They afford a visible goal which may be the end of the journey or simply a stage along the road. In the latter case they help to articulate the route. If a series of landmarks becomes visible as progress is made the driver will tend to form a mental image of the route as being divided into stretches with a particular landmark visible along each.

Landmarks can be reassuring to those not familiar with the route and provide a sense of identification and anticipation to those who know it well. Clearly the most effective landmarks are those which are elevated and isolated. They should have a distinctive shape and stand out against their background or be silhouetted on the skyline. It is often necessary in the aligning of high-speed roads to increase the elevation in order to provide clearance at an underpass or river crossing. Advantage can sometimes be taken of this need to provide an artificial hill by presenting vistas which otherwise would not be seen.

The surroundings in motion

When travelling at constant high speed on a flowing alignment the driver can easily have the illusion that it is the surroundings, rather than his vehicle which are in motion. At the same time everything does not appear to move at the same rate. The closer an object is to the vehicle the faster that object appears to change position. Those very close to the vehicle flash past. As we have seen, it is not possible to focus on them. Elements in the middle distance can be viewed for a short time and those in the far distance can seem to be almost static, **6.7**. Objects close by may appear to move against a static background. Objects to the side of the line of motion will pass parallel to the road edge at apparent

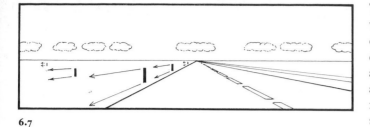

6.7

speeds proportional to their distance from it. Those straight ahead behave in a different fashion. An object in the far distance will, at first, seem to stand motionless and unchanging. Eventually it will appear to grow in size at an accelerating rate as it comes closer and then will suddenly rush past. This illusion of growth is most dramatic for objects close to the side of the road but it is also characteristic of objects some distance from the roadside, **6.8**.

The relative motion of objects to each other and their variations in size depending on their distance from the driver can result in the illusion of an elaborate and graceful dance being performed in the landscape. Objects approach and recede, grow and diminish in size. They draw near and away from each other, reverse position, disappear and then reappear in a new form. A simple diagramatic illustration is shown in **6.9**, where two juxtaposed objects are approached and passed on a straight length of road. It can be seen how the cone draws near, moves behind and then reappears on the other side of the rectangular block. When one element moves behind another in this way dramatic transformations can result. If the object is seen at a distance before disappearing, then reappears in close-up, aspects which had previously been invisible will suddenly become apparent. Especially if the object is of an irregular shape its appearance in close-up may be a transformation of that seen at a distance. As the rectangular block draws near it

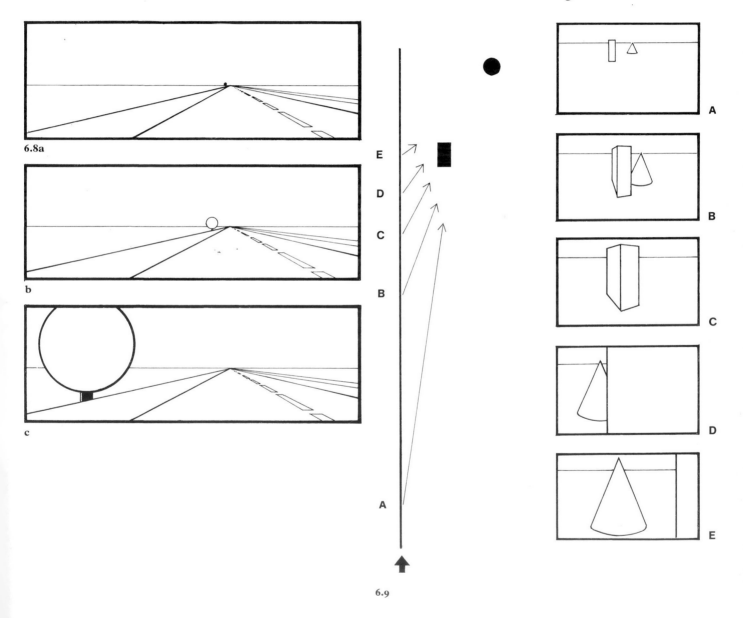

6.8a

b

c

6.9

153

seems to turn on its axis towards the viewer.

Relative movements are more marked when the vehicle is travelling along a curved route, **6.10**. Large factories, power stations or groups of tall buildings in the landscape can exhibit the characteristics described but so too can the natural features of the landscape, the ground formation, vegetation and stretches of water, when, by graceful rearrangements of the constituent elements, caused by the motion of the observer, one view is gradually transformed into another.

The position of juxtaposed objects in relation to the road corridor has a significant effect on the driver's sensation of speed. Those close to the road convey a greater impression of movement than ones further away. Furthermore a series of objects at a given distance from the road and parallel to it will produce a greater impression of speed the closer together the objects are located. Thus if a row of trees is located near the side of the road then the closer the spacing the faster the driver will feel he is travelling.

The gradient too has an effect on the sense of speed. Apparent speed is greater going uphill and less going downhill than on the flat so there is a tendency towards greater speed on the downhill runs.

The road itself can be conceived as moving through the environment. When seen in this way it is necessary that any changes in direction are visually explained if a sense of rightness in the choice of line is to be felt by the user. This will be the case if the road is seen to be responding to changes in topography or to the position of structures and vegetation, **6.11**.

6.10

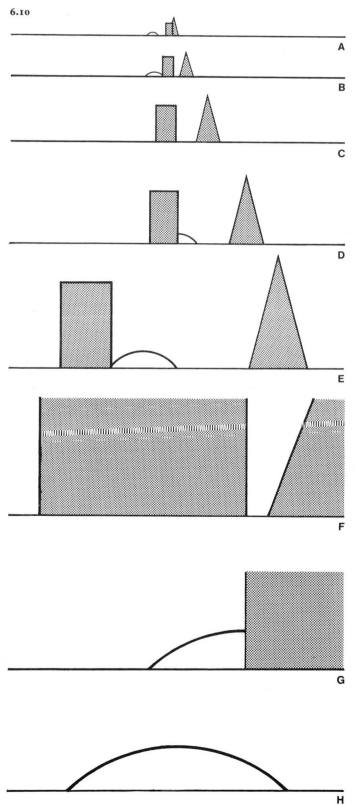

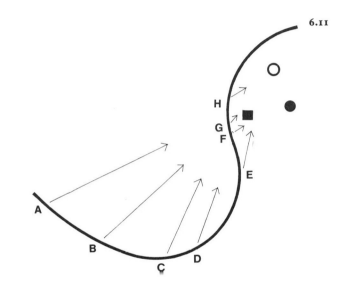

6.11

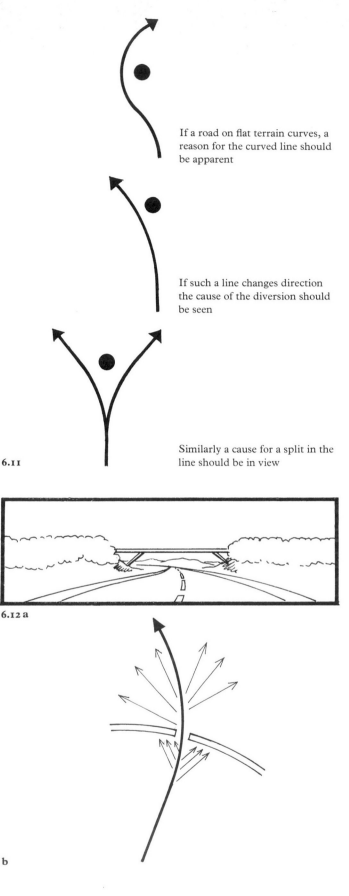

If a road on flat terrain curves, a
reason for the curved line should
be apparent

If such a line changes direction
the cause of the diversion should
be seen

Similarly a cause for a split in the
line should be in view

6.11

6.12 a

b

Contrasts

It should always be the aim of the designer to understand
the character of each of the places being traversed by the
route and to prepare a design which assists the traveller to
experience this character. All aspects of design should be
enrolled to this end. The different degrees and types of
spatial enclosure have their counterpart on high-speed
roads and can be enlisted, as can a number of the elements
of townscape. However we will limit ourselves to the
mention of one of the devices whereby the awareness of the
traveller can be heightened; namely the use of contrasts.

This attribute of a design should not be forced but can be
used to great advantage when the proper conditions
prevail. The contrast between high and low can be used to
emphasize the nature of the landscape. Descending from
high ground in the direction of a lake, a driver's experience
of the change in elevation will be intensified if the road
approaches the water's edge. In traversing a ridge, if the
ground contouring and vegetation on the upgrade closely
contain the road then the open views and sensation of
descent on the other side will be more vividly experienced.
The difference between light and dark when moving from a
moor into a wooded area adds to the excitement of a
journey. The contrast between wide and narrow can be
emphasized by the vertical walls of a rock cutting or a belt
of trees which approaches the road on either side. A notable
event relating to contrasts occurs when the route en-
counters an overpass. The approach embankments to the
overbridge block the distant view on either side of the main
road and after passing through the gap spanned by the
structure a feeling of release is enjoyed on sighting the
uncontained view, **6.12**. A particularly strong impression is
made on the observer by objects which pass overhead, for
example a bridge soffit, **6.13a**, or a tree canopy and those
which come close to the edge of the road, **6.13b**.

As well as enhancing the traveller's awareness of the
character of the places being crossed by the route, a design
should aim to convey a sense of unity to the journey along
a road. This involves solving the problems of sequential
forms: unifying a series of changing and merging fields of
vision in a way which is satisfying to the observer. And so it
is of particular importance that consideration is given to the
mode of transformation from one zone to the next.

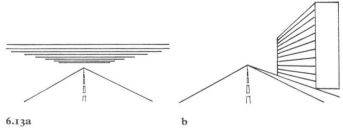

6.13a   b

Transformations

We have mentioned how a series of different views may be presented with one starting after the other has been passed, how such a series of views can overlap each other and how the motion of the vehicle can cause a gradual, continuous transformation of a particular view. There are many other ways in which the zones passed through can be related to each other visually. Some of the possibilities are represented here by simple diagrams and a short description is added indicating the significance of each.

6.14a

(a)   One zone ends and another begins: they are related by contrast and the abruptness of the change. The sudden movement from open countryside to a built-up area or the movement from forest to grassland without a transition zone would exemplify this type.

b

(b)   Here the abrupt change in character is accompanied by a unifying element which continues from one zone into the next. This would pertain if trees lining a route in the open countryside were continued into urban surroundings. The band of the road itself also provides a continuing element.

c

(c)   An element characteristic of a subsequent zone can gradually be introduced in greater and greater quantity to the zone being traversed. This sometimes happens naturally when open ground is followed by scrubland then scattered pioneer trees before a forest is reached. It often happens with much less pleasing results around cities in which the suburbs become more and more dispersed the further they are from the centre until an urban environment becomes a landscape with scattered individual dwellings or clusters of dwellings.

d

(d)   The strong identifying characteristics of one zone are gradually phased out and those of the following zone gradually built up in intensity.

e

(e)   This is the same as (d) but with a common continuous element.

f

(f)   Here two zones with the characteristics of (d) are separated by a shorter one with a different character. In a landscape this could occur, for example, when the primary zones are separated by a lake which is gradually approached then gradually drawn away from by the route, thus creating a highly distinctive intermediate zone.

g

(g)   This combination of (a) and (d) is self-explanatory.

h

(h)   The attributes of the second zone are gradually introduced along the length of the first zone until those of the second become dominant. This type of transition could occur naturally, for example, where slightly undulating ground gradually becomes hilly terrain.

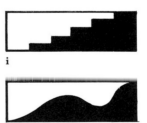

i

j

There is an endless number of ways in which the transition of type (h) can be effected: (i) and (j) suggest two.

PART THREE

LAND FORM AND PLANTING

7
ROADS AND LAND FORM

In a particular design there are many factors which influence the final choice of alignment including the desire lines for vehicle movements, the location of existing structures and developed areas, the existing road networks, the best location for interchanges and the most suitable sites for bridges. Nevertheless every effort should be made to select a route which is sympathetic to the land form.

All the natural features of the landscape exist in harmony with each other, the relation of each part to the whole being dictated by the laws which govern their development. The evolution of each part and its appearance is conditioned by related parts. So the river is in harmony with the valley; the rock ouctrop with the hillside; the plant communities with the local soil types and topography. Similarly roads should be formed as an inter-reaction between the route and the landscape; the former being moulded by the latter. they should look as inevitable and as much a part of the whole as the elements of the natural environment.

The route and the ground contours

A good road alignment will take advantage of the natural land form. The line will merge with the existing ground contours rather than attack them, **7.1**.

Routes running along contours blend most easily with the landscape and are least demanding on vehicles and pedestrians alike, **7.2**.

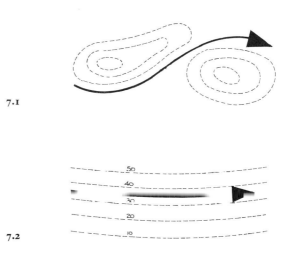

7.1

7.2

Buildings which line roads laid parallel to the contour lines sit easily in the landscape. When the ground slopes steeply such structures are displayed to good advantage, as is the case with the terraces in this view of Bath, **7.3**. In open ground a well-sited road parallel to the contours will emphasize the land forms, **7.4**.

7·3

7·4

When it is necessary to increase or reduce the elevation along the length of a road which is running in the direction of the contours on sloping ground, a comfortable gradient can be fixed by selecting an appropriate angle between the road line and the contour lines **7.5**.

A road climbing in this way can result in prominent scars, **7.6**, which must be avoided by careful selection of the exact line and gradient and by careful detailing of the slopes of cut and fill.

When a route runs at right angles to the contours, **7.7**, a location should be chosen which minimizes the apparent amounts of cut and fill. The appearance of a route forced through the landscape 'against the grain' should be avoided. This may be achieved by choosing a line and levels which minimize the actual quantities of cut and fill. On the other hand some deviation from this position may produce the best results.

A line at right angles to the contours can have a strong visual impact so, here too, particular care must be given to the treatment of side slopes in cut and fill. This is especially true on steeply sloping land where an ugly gash in the terrain can be the result of unsympathetic treatment, but it is also true of relatively flat, undulating land, **7.8**.

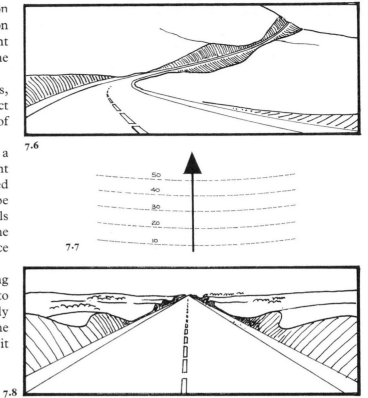

7.6

7.7

7.8

7·5

When it is necessary to align roads at right angles to the contours in built-up areas problems can arise in relating buildings to the road on steeply sloping ground. One way that difficulties can sometimes be overcome is to ensure that the buildings do not face onto the road with the steep gradient but onto roads at right angles to it and thus parallel to the contours. However if buildings do face onto the steep road the results can be dramatic if the difficulties are successfully overcome, as in the design of photograph **7.9**.

7·9

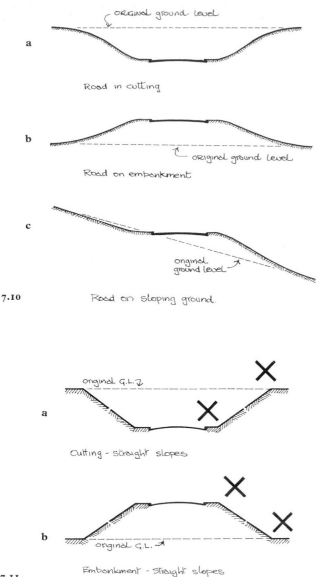

a original ground level

Road in cutting

b original ground level

Road on embankment

c original ground level

Road on sloping ground.

7.10

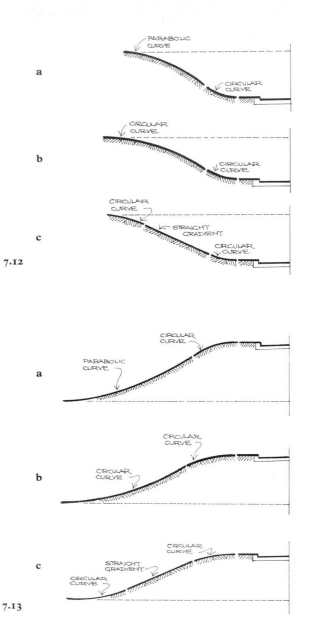

a PARABOLIC CURVE / CIRCULAR CURVE

b CIRCULAR CURVE / CIRCULAR CURVE

c CIRCULAR CURVE / STRAIGHT GRADIENT / CIRCULAR CURVE

7.12

a original G.L. / Cutting - straight slopes

b original G.L. / Embankment - straight slopes

7.11

a PARABOLIC CURVE / CIRCULAR CURVE

b CIRCULAR CURVE / CIRCULAR CURVE

c STRAIGHT GRADIENT / CIRCULAR CURVE / CIRCULAR CURVE

7.13

The cross-section

When roads are in a cutting the ground should be modelled on either side to make their position appear as natural as possible, **7.10a**. The remodelled ground should have a similar character to that of its surroundings. This is also true when the road is on an embankment, **7.10b**, or when it runs in the direction of the contours on sloping ground, **7.10c**.

Steep sided slopes and squared shoulders and bases should be avoided both for cuttings and embankments, **7.11a** and **b**.

An exception is the case of a cutting in rock where a steep face and sharp angles are natural to the material.

The precise way in which the side slopes of cuttings and embankments are to be profiled can only be decided properly after careful study of the surrounding ground formations. The general land forms in the area should be noted, as should those in the immediate vicinity of the length of the proposed road and the shape of the earthworks should be designed accordingly. Normally the side slopes should be made as flat as possible. The land surface has been formed by erosion through the action of water and except in the case of rock outcrops the surfaces will be made up of rounded forms and rounded junctions of planes.

With cuttings excellent results can be obtained by making the sides convex. Two ways of doing this are indicated in **7.12a** and **b**.

An alternative method is to use a side gradient which is rounded off at the top and the bottom, **7.12c**. Or it can sometimes be continued to the road edge, **7.14**.

7.14

7.15

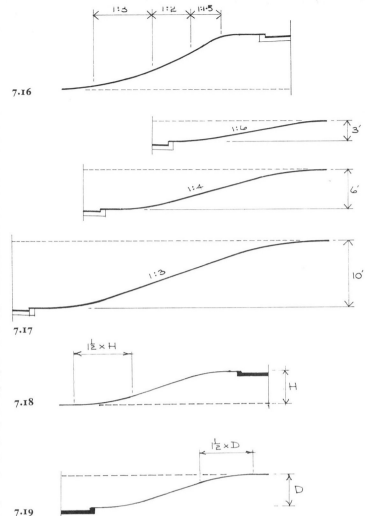

7.16

7.17

7.18

7.19

The side slopes of embankments will invariably look best if the sides are predominantly concave. This can be achieved by making the upper part of the slope steeper and the lower part flatter than the average, so that the line of the slope takes the shape of a parabolic curve with the upper part of the slope rounded off, as in **7.13a**. Another way of achieving a similar effect is to use circular curves with a longer one at the bottom than at the top, **7.13b**. A flat slope with the gradient rounded off at the top and the bottom is another possibility, **7.13c**. That flat side slopes curving into the existing terrain are visually pleasing is well demonstrated by photograph **7.15**.

The steepest side slope which can be used is determined by the engineering properties of the soil. Usually this will permit a gradient which is steeper than that giving the best visual effect.

Flattened and liberally rounded cut and fill slopes, in character with the topography, are recommended in *A Policy on Geometric Design of Rural Highways*.[16] Proper shaping of the side slopes, it is stated, is necessary to obtain effective erosion control, low cost maintenance and adequate drainage of the subgrade. It is pointed out that maintenance equipment is most efficient on 1:3 and flatter slopes and that 1:4 and flatter slopes reduce the risk of a serious accident in the case of a vehicle leaving the carriageway. It is easier for the driver to maintain control and impact is lessened on cut slopes. An additional advantage of flat side slopes in cuttings is the longer sight line obtained. Well-rounded forms produce a stream-lined cross-section which reduces wind erosion and snow drifting. It is considered that the best results are obtained by the individual study of each case, the use of contour maps being particularly helpful.

Ritter and Paquette, in *Highway Engineering*,[17] point out that in the United States 1:4 side slopes for both cut and fill are used a great deal to a depth or height of about 10 ft and when the cut or fill does not exceed 6 ft they recommend a maximum side slope of 6:1.

In *Man-made America*,[18] it is suggested that for embankments up to 12 ft high or even more a slope as flat as 1:4 or 1:6 should be used. The extra cost of earthworks can be offset against the saving in cost of a visually objectionable crash barrier.

In *The Landscape of Roads*,[19] Sylvia Crowe gives the average side slopes which were used in the construction of the Ulm/Baden-Baden autobahn, shown here in **7.16**.

The paper by Cordell and Howarth,[20] who recommended that the lower a cutting the flatter should be the slope, is quoted and she points out that this will clearly increase the impression of an easy flow between the road and the countryside. The relationship between depth of cut and the gradient of the side slope as recommended by Cordell and Howarth is indicated in **7.17**.

It is common practice in the United States to vary the side slope with the height of the embankment or the depth of the cutting. *The Highway Engineering Handbook*[21] lists the slopes adopted in the construction of the Garden State Parkway, New Jersey. The figures are reproduced below. Further examples of American practice are listed by Tunnard and Pushkaref.[22]

Depth (ft)	Cut slopes	Height (ft)	Fill slopes
0–3	1:6	0–5	1:6
3–6	1:4	5–12	1:4
6–25	1:3	15–50	1:2

In the AASHO's *A policy on landscape development for the national system of interstate and defence highways*[23] it is suggested that the roundings at the bottom of a side slope in fill should have a length 1½ times the height of the fill. In cutting, the rounding at the intersection of the side slope and the original ground plane should be 1½ times the depth of the cutting, **7.18** and **7.19**.

The use of closely spaced contour lines on the plan of the new roadworks can facilitate the work of both the designer and the contractor. In *The Highway Engineering Handbook*[24] the writer observes that although most highways are constructed from drawings showing the profile of the centre line and cross-sections at 100 ft intervals, many of the American parkways and expressways have been designed throughout with grading plans showing contours at 1 or 2 ft intervals. This, it is asserted, is the best way of assuring the desired formations of earthworks and producing good appearance, good surface drainage and facilitating the calculation of earthwork volumes as construction progresses.

Whilst recognizing that American practice is not necessarily directly applicable in the UK owing to differences in land use and cost, nevertheless the trend towards flatter side slopes in the United States should encourage UK designers to consider carefully the benefits of this choice when designing their own roadworks.

When a road which is aligned parallel to the contours is in cutting it sometimes happens that a small nib of ground would remain after the necessary amount of material had been removed. Consideration should be given in the design to the possibility of removing this material in order to avoid an awkward appearance in the final land forms, **7.20a** and **b**.

Similarly when an embankment is being designed on sloping ground it may be feasible to better integrate the earthworks by filling in the volume between the embankment and the original ground on the high level side, **7.21a** and **b**. Of course, precautions must be taken to ensure that water from the high land does not drain onto the road surface or saturate the base or sub-base.

In determining the design of side slopes, maintenance and future land use should be considered. Gang mowers and most other grass cutting equipment cannot be conveniently used on slopes steeper than 1:3. By employing flat slopes land which would otherwise be unused can be grazed or cultivated. Slopes of 1:3 can be grazed, and the minimum slope necessary for regular ploughing is 1:6 or 1:7. Figures **7.22a** and **b** illustrate how land can be saved by using flat side slopes either in cutting or on embankment.

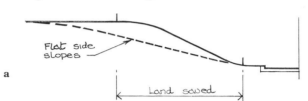

a

Cutting – Alternative side slopes

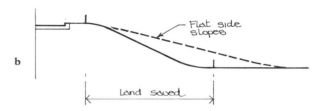

b

7.22 Embankment – Alternative side slopes

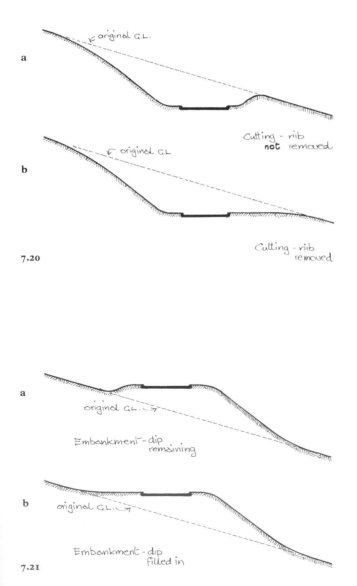

7.20

7.21

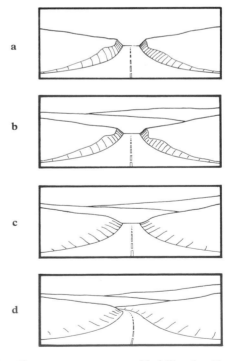

a

b

c

d

7.23

Straight cuttings cause an unnatural-looking break in the land form, especially when seen on the skyline, **7.23a**.
The cut will be less obtrusive if seen against a background rather than in silhouette, **7.23b**.

Modelling the sides greatly helps to make the line a part of the landscape, **7.23c**. The best result is likely to be attained if the line can be so arranged that the cutting is on a horizontal curve, **7.23d**.

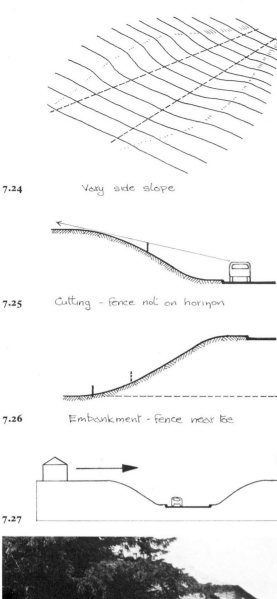

7.24   Vary side slope

7.25   Cutting - fence not on horizon

7.26   Embankment - fence near toe

7.27

When the length of road is in cutting its depth will vary as the road enters and leaves the cutting and also within it unless the ground contours are very regular. A more natural appearance will result if the side gradients are not kept constant but vary with the depth, the deeper sections of the cutting having steeper side slopes than the shallow ones. It is possible to do this by keeping the width of the cutting constant and varying the angle of the slope. This, however is likely to be unsatisfactory since keeping the shoulder parallel to the road accentuates the parallelism of the road. It is better to model the ground in such a way that the shoulder curves in towards the road as the depth of cutting decreases, 7.24.

The position of fences on the cross-section should be considered with care. In cuttings it is advisable not to put them at the top of the slope where they will be seen in silhouette from the road. They are best situated below the skyline where they will be less conspicuous, 7.25, and if aligned parallel with the top of the cutting they are most likely to be in sympathy with the land form. An alternative position is beyond the shoulder where they are not visible from the road but this is likely to be wasteful of land. Fences are usually placed at the bottom of embankments and this is normally the best place for them, 7.26, if flat side slopes are not being used. Where flat side slopes are used near the bottom of the embankment it may be possible to save some land by placing them part way up the slope.

Lowering the alignment below the existing ground level can greatly reduce the impact of the road on the surrounding environment, 7.27. As well as protecting the views from vantage points outside the road this is an excellent way of containing traffic noise.

7.28

An urban example of the road being located at a lower level than the adjacent dwellings is shown in 7.28.

In some instances it will be desirable to conceal the road although a deep cutting is not necessary or practical. In such a case it will sometimes be possible to locate the road at a level slightly below that of the surrounding ground and use the excavated material to form an embankment at the side, as illustrated in figure **7.29**.

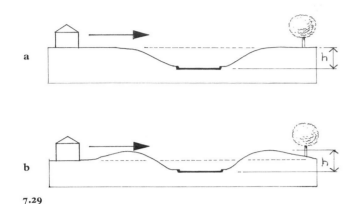

7.29

7.31

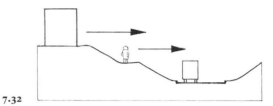

7.32

7.30

7.33

Even small differences between the road level and the general ground level can render the road less conspicuous. The low rounded banks prevent cars being parked off the road surface on the verge, **7.30**.

Lowering the road a small amount below the general ground level in a residential area can be used to differentiate between the domain of the vehicle and that of the pedestrian, **7.31**. This expedient is particularly useful when the excavated material is required as fill elsewhere on the site. If this is not the case it is often possible to arrange the landscaping to utilize any surplus material. Even in areas where mature trees grow close to the road edge the lowering of the surface can sometimes be carried out provided the tree roots are carefully pruned and proper provision is made to maintain the original ground level in the vicinity of the trees, **7.30**.

Lowering the road in relation to local dwellings and footpaths can give the occupants and pedestrians a view over the tops of parked and moving cars **7.32** and **7.33**. If pedestrian routes are required parallel to the road they will best be situated at the top rather than the bottom of a cut. This will afford a much pleasanter and safer environment. The same considerations apply to cycle tracks.

An example of the footpath being situated at the top of the cutting for a local distributor is shown in **7.34**.

In the countryside or an Arcadian setting when there is a flat side slope to the cutting the footpath can be kept some distance from the road and this distance can be varied along the route, **7.35**. The principle of taking advantage of the side slope to raise the footpath above the level of the road can also be applied in a more urban context. The level of the pedestrians on one side of the village street in **7.36** has been raised. The resulting contrast between the two sides greatly contributes to the visual delight of the whole. The residential road in the Arcadian environment of photograph **7.37** separates the pedestrian from the vehicle in a similar manner. Steep side slopes have been used as can be seen on the far side of the road in the photograph and these, combined with the hedging and the canopy of trees, provide a strong sense of enclosure to the area and the two tier segregation system contributes valuable visual interest.

Even small differences in level between footpath and road can greatly enhance the position of those on foot. This is especially so if the verge formed by the bank is used for tree planting, as in **7.38**.

7·34

7·35

7·36

7·37

7·38

167

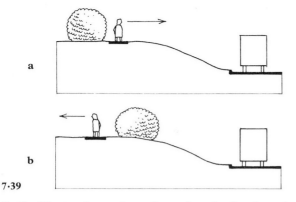

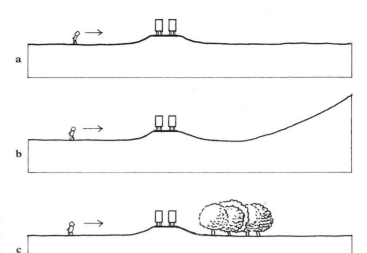

7.39

In the illustrations where the pedestrian has been located at the top of the bank he has been afforded views across the road. An alternative arrangement is to screen the footpath from the road and provide views in the opposite direction, **7.39**.

7.41

Roads on embankments pose their own special problems with regard to their integration into the environment, and the higher the embankment the more difficult the problems become. Nevertheless when it is necessary to elevate the road in this way an opportunity is provided to open up views and so increase the interest of the route from the point of view of the road user. An example of full use being made of this opportunity is shown in photograph **7.40** of a district distributor road.

Embankments are most conspicuous when seen in silhouette, **7.41a**. They will be better integrated into their surroundings if they are positioned to be seen against a background of higher land **7.41b**. If the higher ground is close by, as shown in the sketch, a better location for the road would possibly be on the side of the hill. If no high land forms a backcloth to the embankment then planting can be used for this purpose, as illustrated in **7.41c** and **7.42**.

7.42

7.40

When necessary the presence of an embankment can be screened by planting. A row of trees along the side can help to make the road a part of the landscape. This is shown in photograph **7.43**, where the road is traversing a small stretch of common land in an urban area.

An embankment can also have the effect of imparting a sense of enclosure to the area on either side. Of course, the higher the bank the more this effect will be in evidence but even the slight raising of the ground in **7.43** helps to convey a sense of containment to the foreground area.

When footpaths are required parallel to a road which is on an embankment, as was seen to be the case with cuttings, the best position is not necessarily that adjoining the road. In this location the pedestrian will have maximum exposure to the sight, noise and fumes of the traffic, **7.44a**. Three alternative locations are shown in **7.44b**, **c** and **d**. A footpath screened from the traffic but on top of the embankment is likely to afford the best views. On high embankments a position half way up the side may be preferable since some screening from noise will be obtained. The most sheltered position with maximum screening from noise will result from locating the footpath at the bottom of the bank. This will often mean forfeiting the views. As is usually the case, the same considerations apply to cycle tracks as to footpaths.

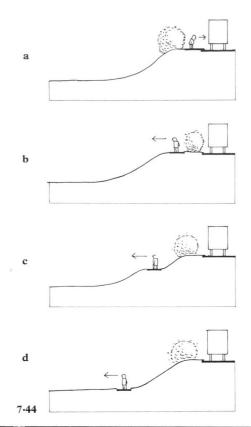

a

b

c

d

7.44

7.43

In cuttings and on sloping ground problems of changes in level may be solved by the use of retaining walls. This is a valuable method particularly in urban situations where the land necessary for embankments is not available or where it is desired to maintain a tight sense of enclosure. Photograph 7.45 shows the adoption of a retaining wall solution in an urban situation, the effect being to separate the dwellings and the access road from the through road. In photograph 7.46 a low retaining wall has been used to raise the footpath above the level of the road.

In the urban context, if a residential area is located at a lower level than the adjacent road the pedestrian-only precinct is clearly defined, 7.47. It can be seen that having the road higher than the general ground level of the residential area can sometimes make the road less prominent. Of course, other elements of the urban environment must be recruited towards achieving this effect.

7·45

7·47

7·46

Mounding

In addition to linear cuttings and embankments there is a third way in which the designer can alter the land form to achieve his ends. This is by the use of mounding. Mounding entails the construction of volumes of earth above original ground level. These are modelled into a shape which fulfils some purpose in a design. Whether used on a large or a small scale great care must be taken to ensure that the shapes used are in character with the surrounding landscape and the other aspects of the design.

7.48 shows the use of mounding to integrate a slip road, the approaches to an overbridge and a section of dual carriageway into the land form and to act as a unifying element in the total design.

7.48

7.49

7.50

7.51

The high continuous mound of **7.49** has been used to separate the residential area and the access road from the busy district distributor on the other side.

The ground contours in **7.50** have been arranged to form an embankment between the pedestrian route and a local distributor road.

All sides of the road in **7.51** are contained by mounding and a barrier is thus placed between the road and the dwellings. The tree planting has been carefully arranged in relation to the land shapes to form a containing screen in the background.

Linear mounding together with trees in **7.52** ensure a pedestrian route out of sight of the road.

In photograph **7.53** the ground enclosed by a small loop road serving a group of dwellings has been raised and modelled, the purpose being to break up the view of the road surface which would otherwise have been too prominent and to add interest to the space enclosed by the dwellings.

The slight mounding on the side of the access road in **7.54** adds to the enclosure of the footpath on the other side of the birch trees. A detail such as this enhances our environment out of all proportion to the cost of creating it.

7.52

7·53

7·54

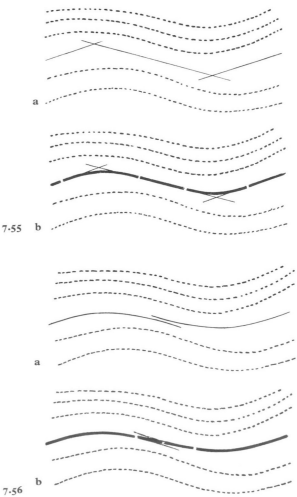

a

7·55 b

a

7·56 b

Drawing the alignment

The technique adopted for drawing the road line can have a marked effect on the final construction. The traditional method of producing the plan has been to draw down the tangent lines on a contour map, **7.55a**, and then to fit horizontal curves and straights to this framework, **7.55b**. An alternative to this procedure, which can produce a more flowing appearance, is to work with curves from the start of the drawing process. Thus a series of curves are fitted to the contours, **7.56a**, in such a way that they can be easily joined by transition curves, **7.56b**.

Roads and land form—Summary

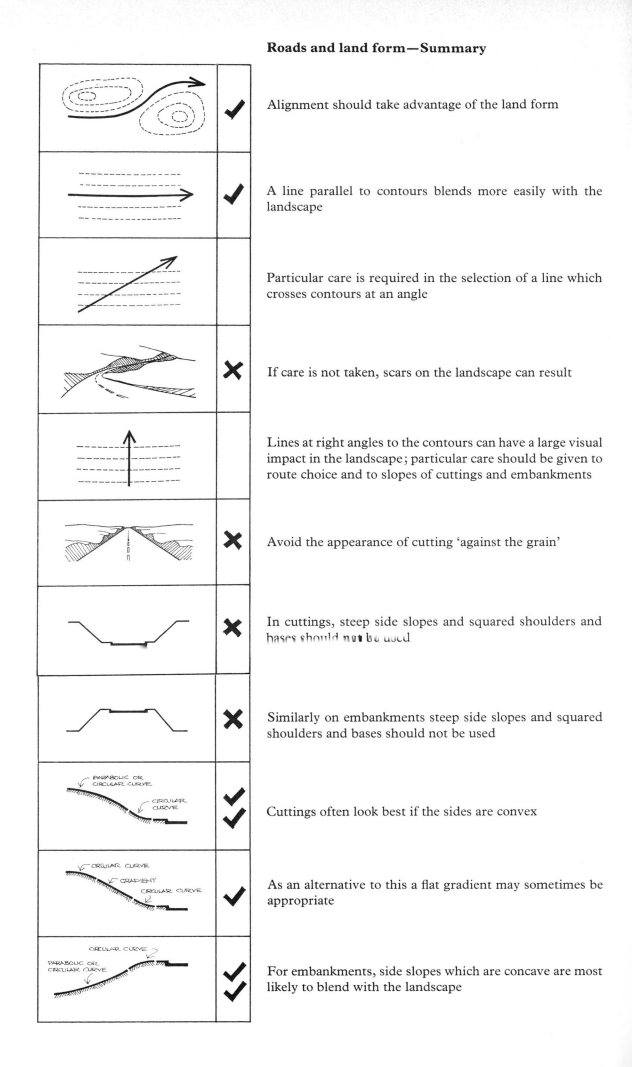

Alignment should take advantage of the land form

A line parallel to contours blends more easily with the landscape

Particular care is required in the selection of a line which crosses contours at an angle

If care is not taken, scars on the landscape can result

Lines at right angles to the contours can have a large visual impact in the landscape; particular care should be given to route choice and to slopes of cuttings and embankments

Avoid the appearance of cutting 'against the grain'

In cuttings, steep side slopes and squared shoulders and bases should not be used

Similarly on embankments steep side slopes and squared shoulders and bases should not be used

Cuttings often look best if the sides are convex

As an alternative to this a flat gradient may sometimes be appropriate

For embankments, side slopes which are concave are most likely to blend with the landscape

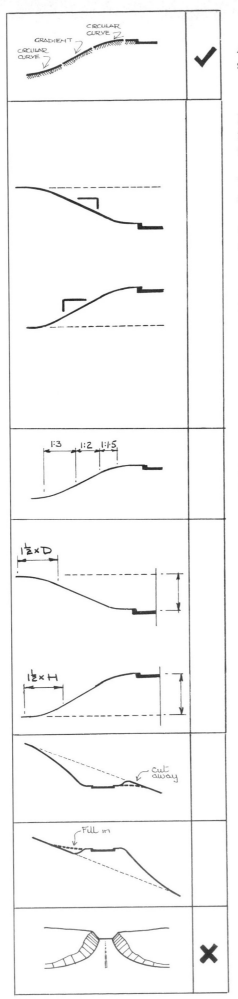

✔ Alternatively a flat gradient may sometimes give the best shape

Ritter and Paquette say 1 :4 side slopes much used in cut and fill up to height of 10ft; when cut or fill does not exceed 6ft they recommend a maximum side slope of 1 : 6

Tunnard and Pushkarev recommend 1:4 or 1:6 for embankments up to 12ft

Cordell and Howard recommend for cuttings 1:6 for 3ft depth, 1:4 for 6ft and 1:3 for 10ft

Slopes used for the Arden State Parkway, New Jersey, listed in table below

Depth (ft)	Cut slopes	Height (ft)	Fill slopes
0–3	1:6	0–5	1:6
3–6	1:4	5–12	1:4
6–25	1:3	15–50	1:2

Embankment side slopes used for Ulm/Baden-Baden autobahn

In *A policy on landscape development for the national system of interstate and defence highways* AASHO suggest rounded lengths as shown

In case of nib remaining, consideration should be given to its removal as shown

In case of embankment on side slope, consideration should be given to filling in as shown

✘ Straight cuttings viewed in silhouette should be avoided

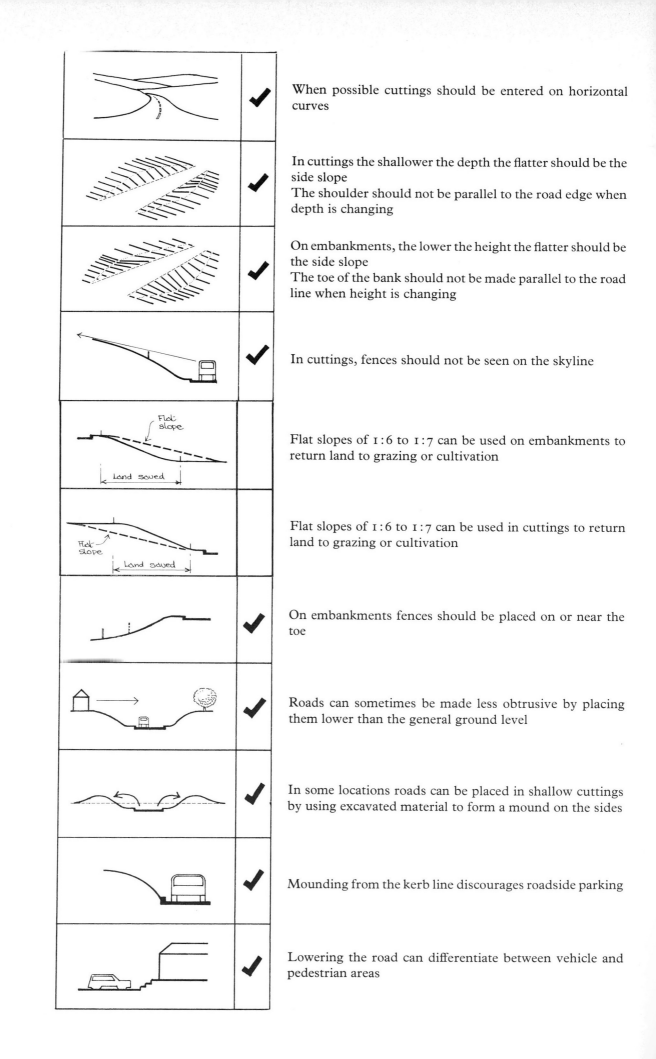

When possible cuttings should be entered on horizontal curves

In cuttings the shallower the depth the flatter should be the side slope
The shoulder should not be parallel to the road edge when depth is changing

On embankments, the lower the height the flatter should be the side slope
The toe of the bank should not be made parallel to the road line when height is changing

In cuttings, fences should not be seen on the skyline

Flat slopes of 1:6 to 1:7 can be used on embankments to return land to grazing or cultivation

Flat slopes of 1:6 to 1:7 can be used in cuttings to return land to grazing or cultivation

On embankments fences should be placed on or near the toe

Roads can sometimes be made less obtrusive by placing them lower than the general ground level

In some locations roads can be placed in shallow cuttings by using excavated material to form a mound on the sides

Mounding from the kerb line discourages roadside parking

Lowering the road can differentiate between vehicle and pedestrian areas

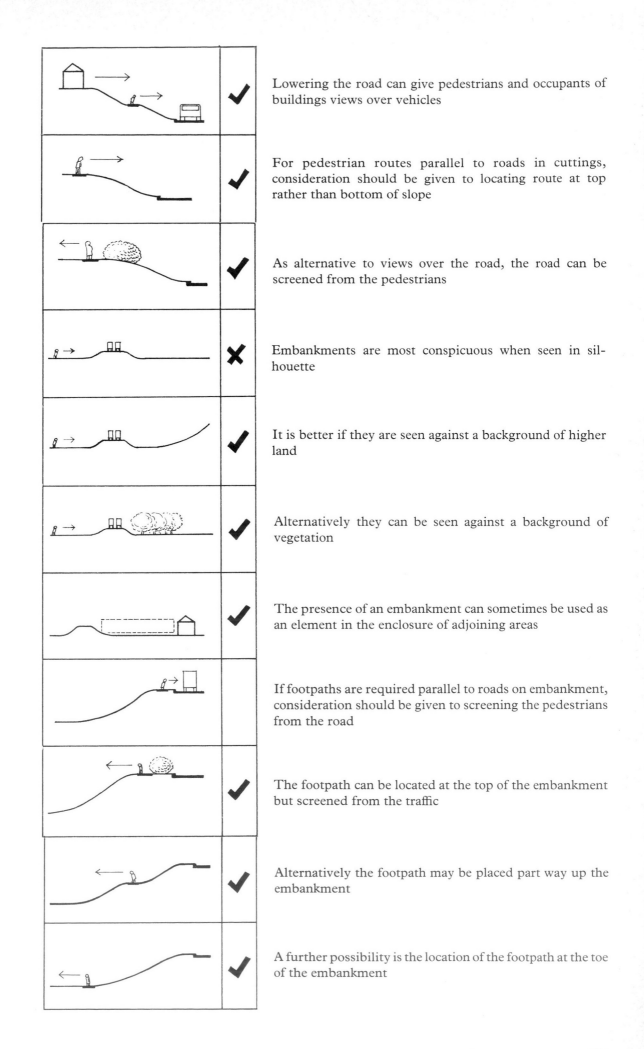

Lowering the road can give pedestrians and occupants of buildings views over vehicles

For pedestrian routes parallel to roads in cuttings, consideration should be given to locating route at top rather than bottom of slope

As alternative to views over the road, the road can be screened from the pedestrians

Embankments are most conspicuous when seen in silhouette

It is better if they are seen against a background of higher land

Alternatively they can be seen against a background of vegetation

The presence of an embankment can sometimes be used as an element in the enclosure of adjoining areas

If footpaths are required parallel to roads on embankment, consideration should be given to screening the pedestrians from the road

The footpath can be located at the top of the embankment but screened from the traffic

Alternatively the footpath may be placed part way up the embankment

A further possibility is the location of the footpath at the toe of the embankment

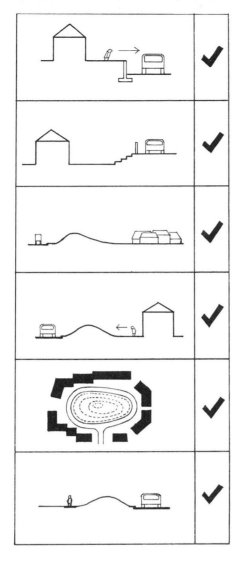

In urban situations changes in level can be accomplished by the use of a retaining wall

Placing a residential area at a lower level than the road can sometimes be used to separate the pedestrian and vehicle territories

Mounding can be used to separate a residential area from the main road

The separation of vehicles from single dwellings or groups of dwellings can be brought about by the use of mounding

The visual impact of a road can sometimes be reduced by the use of mounding

Mounding can be used to help screen a footpath from the road

8
ROADS AND PLANTING

The presence of plants in the environment satisfies a deep psychological need in all of us. In the past they have always formed a vital part of man's surroundings and today even the city dweller is seldom far from the sight of a tree.

Plants have many uses in relation to roads. They can fulfil engineering functions; the purification of the air by absorbing noxious gases, emitting oxygen and filtering out dust is one such service which they perform. With their roots they stabilize side slopes against erosion from wind and rain. They can be used to combat glare from oncoming headlamps and to screen traffic from surrounding buildings. Noise control and traffic guidance also number among their possible uses. Climate control too may be achieved through the provision of shade and shelter from the wind or rain. Tree belts can be used to protect roads from the worst onslaughts of prevailing winds, **8.1**. Plants can be used to control the wind issuing from draughty gaps between buildings, **8.2**.

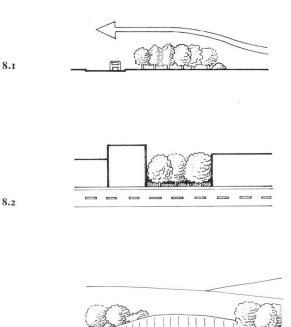

8.1

8.2

8.3

Particularly disturbing and sometimes dangerous for the motorist is when a vehicle is suddenly struck by a strong wind on emerging from a cutting. Judicious tree planting can ensure that the build-up of wind force on the vehicle is a gradual one to which the driver can adjust, **8.3**.

Organic materials can be employed for their architectural properties. Grass verges may form part of the floor of the space containing the road; the walls can be of trees and shrubs and a canopy form the ceiling. Plants may be employed as screens, as means of orchestrating views, for the articulation of space and the control of scale.

Aesthetically, the attributes of plants and their possibilities in the hands of a designer are endless. They can be used for the creation of pattern and texture, for the calligraphic effects of their winter silhouettes, for their sculptural shapes and as decorative walls. In relation to buildings they can form a green background, a foreground frame or simply an object whose organic structure and form complements the geometric properties of a building.

Many of these matters have been treated in some detail elsewhere, and a list of relevant references is presented in the bibliography at the end of the book. However, this book is concerned with road form and townscape so the consideration of planting will be confined mainly to some of the architectural and strategic uses of vegetation.

Planting and landscape

RELATION TO COUNTRYSIDE VEGETATION

The primary aim in the use of planting in relation to roads in the countryside is the integration of the road into the landscape. This must be done on a scale which is in accord with the plant growth of the area, with the land forms and with the speed of traffic on the road. The designer must be conscious of the need to enhance the views from outside the road corridor as well as the constantly changing view of the road user.

For the driver looking straight ahead the context in which the road is seen will be a landscape in a continuous state of transformation. On straight lengths of road the view ahead will grow and change as the part of the countryside within the cone of vision alters; on curved lengths of road the field of vision will change as the curve is traversed. For those looking from the side of the car the way in which the views evolve is different. The rate of change will depend on the nearness of the view as well as the speed of the vehicles. Distant vistas will be apparently static for quite long periods; near objects will flash past. Plants along the road corridor should be designed to the scale of the vehicle speed. Belts of trees bordering high speed roads will be elongated if they are to register on the consciousness of the driver. Similarly vistas which are merely glimpsed and then disappear behind a wall of foliage will be an irritant, but viewed for a longer period they will enhance the journey. Thus gaps between the tree belts also must be scaled to the speed of travel.

The vegetation of any stretch of countryside being traversed by a road will consist of plants which are suited to the soil type and climate. Their arrangement on the land will be a function of the topography, natural ecological groupings and the ways in which they have been influenced by man's activities. The designer must study all these factors. Every effort should be made to build new planting into the structure of the existing vegetation. It follows that plant communities should be established which are characteristic of the local ecology and that plant communities and groupings are more important than individual species, although these too are important. The new planting should complement that which is already in existence. It should elucidate and sometimes emphasize the land forms as well as being an integrating element for the road. There is no place for decorative planting in the open countryside.

In wooded areas the road will seem less of an intrusion if the tree line is not too distant from the road edge and if the edge of the wood is planted with those trees and shrubs which normally occupy a position on the woodland perimeter. Yet however natural the road alignment appears, a long drive through woodland can become oppressive, as can continuous tree belts planted along the roadside. The driving experience will be greatly enhanced if the confined space of

the woodland corridor is relieved by open spaces and views out. Careful alignment and associated planting can impart a rhythm to the visual experience of the driver by the provision of a series of open and confined situations.

This rhythm will be largely conditioned by the content of the existing landscape. Heavy planting may be required near the road when it passes close to built-up areas, especially when these areas contain housing, to protect them from the sight and noise of traffic. On the other hand planting may be required to screen from the sight of road users areas of industrial dereliction, open cast mining, electrical pylons and other unsightly intrusions on the countryside. At the same time many manifestations of industrial activity can be of great visual interest and nearby or distant power stations, cooling towers and chemical plants may well be left in full view or allowed to be seen in relation to judicious planting.

Belts of trees planted along the road edge should relate visually to established planting in the area. At the same time vegetation in the countryside should usually be placed with a view to reducing the parallelism of the carriageways. Thus if the trees form wooded areas which cut across the line of the road they will tend to reduce this effect and help to make the road a part of the landscape. This is especially true for the views from outside the road boundary. In some cases it may be possible to plant a belt of trees along the roadside which will form implied links, though not necessarily actual ones, with other belts or clumps in the landscape and thus effect the required integration. In planning a road the designer should always consider the possibility of achieving planting in areas outside the road corridor in order to improve both the overall appearance and individual views. This may be possible by agreements with local authorities, landowners, farmers, the Forestry Commission and other organisations with jurisdiction over the countryside. Whether or not this proves possible, the intersections on major roads provide large open spaces within the boundaries of the roadworks where the planting of tree belts can be considered.

The relationship between speed and scale is significant in all aspects of road design. The faster the speed, the wider the lanes, the flatter the curves, the longer the sight lines and the further back from the road edge will be the trees. Nevertheless it is often preferable to allow the planting to approach as close to the running surface as considerations of safety will permit. The more minor the road the closer the trees and shrubs can be expected to approach to the running surface. The trees can be kept back sufficiently far so that their leaves and branches will not fall onto the running surface, **8.4**.

The slower the design speed, the less the danger of serious accidents from vehicles leaving the carriageway and striking the tree trunks. In this regard a shrub layer in front of the trees has the additional advantage of providing a protective barrier. The best shrubs for this purpose will have a dense springy growth. In the United States *Rosa multiflora* has proved effective for this purpose.

The natural climax vegetation over most of the British Isles is woodland. Verges and side slopes of embankments and cuttings, if left to develop naturally, will be invaded by herbaceous plants followed by shrubs then by pioneer species of trees including birch, alder, rowan and sycamore

8.4

8.5

and finally the large forest trees will attempt to establish themselves.

This process can be observed taking place on railway embankments. Those in charge of the design and maintenance must decide for each length of road what is the most appropriate stage at which the natural evolution of plant life should be arrested or modified. Having made this decision a maintenance programme can be worked out to preserve the desired stage of development. In the countryside the key to the correct solution is the character of the surroundings. There are many locations where the side slopes can be left to develop naturally. In other places the best answer may be a grassed slope with clover and wild flowers, **8.5**. Decisions of this kind should be made at the outset since they may affect the desirable contours of the embankments and cuttings. Nan Fairbrother, in *New Lives, New Landscapes*,[25] points out that indiscriminate mowing and spraying is no longer carried out universally and suggests that the best general treatment on verges may be to cut the grass once a year after the wild flowers have seeded raking the hay off to protect the more delicate plants.

8.6

Roads with vegetation on the side slopes and verges are invaluable ecological corridors which form the habitat of many species of birds, small animals, insects and plants. Michael Porter points out in his paper *Some Aspects of Road Landscape*[26] that there are 200,000 acres of roadside verge in England and Wales and that the potential of these second-string nature reserves is immense. This is certainly the case and it is particularly important that this potential be realized since so much of our flora and fauna are being expelled from other parts of the countryside as a result of modern farming techniques, including the use of herbicides and insecticides and the destruction of the hedgerows.

In the treatment of the side slopes and verges as elsewhere, the need to provide a varied and interesting view to the driver is of great importance. Thus it may be desirable to vary the appearance over particular stretches of road. If the general policy is to let the grass grow up and die off naturally with the seasons then in special areas, for example where the road passes near to a housing neighbourhood, the degree of maintenance could be intensified and the grass might be kept trim by frequent cutting.

In areas where the road crosses a landscape of hedged-in fields this pattern should, where possible, be respected in the road design. The same type of hedging can perhaps be used instead of roadside fencing over particular lengths. Where livestock is to be contained a temporary fence will be necessary while the hedge is becoming established. Nevertheless, depending on the lie of the land, parallel hedgerows over large lengths of road can be frustrating for the motorist by continuously blocking the view, so this practice should be used with discrimination. On some minor roads through arable land it has proved possible to dispense with any barrier containing the field. This results in a very strong sense of unity between the road and the land.

Rather than allow natural plant communities to establish themselves another possibility is to design and landscape the view along the road in a studied and semi-formal fashion. This is obviously an expensive undertaking but it is justified in some instances over short lengths of road which call for a special effect. An example of this practice is shown in **8.6** where the landscaping of the roadside views

signals the entry of the traveller into the new town of Welwyn Garden City.

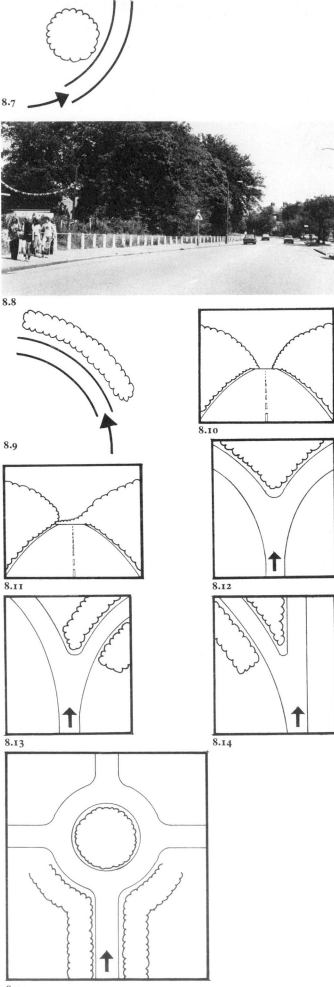

STRATEGIC USES OF PLANTING

We have seen that new planting associated with the road design must relate to the existing countryside vegetation. At the same time it can be arranged to clarify and reinforce the visual logic of the alignment. Some ways in which this can be done are presented below.

As with all other aspects of road design, new planting must be considered at all stages of the design process as a three-dimensional undertaking: the fact that some of the diagrams are presented in plan or elevation arises from the need to isolate particular points from their total context in order to emphasize them and from the convenience of this method of presentation.

A visual logic can be imparted to a horizontal curve, when the reason for it is not immediately apparent, by tree or shrub planting on the inside of a bend, **8.7**.

In **8.8** the curve of the road follows the boundary of some playing fields and the logic of the line is emphasized by the need to avoid the clump of trees adjacent to the inside radius.

Planting on the outside of the curve will emphasize the line and at the same time can seem to dictate the route chosen, **8.9**. This type of planting can be particularly useful for night driving.

The continuation of a line on the other side of a vertical curve can be previewed by the presence of trees at the side of the road.

When there is no change in horizontal direction across the summit curve this is clarified for the driver, **8.10**, and, more important, when there is a change of direction he is forewarned, **8.11**.

The existence of a fork in the road can be emphasized by planting, **8.12**. This is not a desirable form of junction when vehicles are travelling at other than slow speeds. Even then it should only be used after careful consideration of all the traffic implications and possibly after introducing modifications such as a special stopping area for vehicles taking the right-hand fork. Nevertheless there are places where such junctions already exist and there may be other locations where their use cannot be avoided or even where they are desirable for visual reasons and at the same time are feasible as a consequence of slow design speeds. In these cases it is desirable to signal the layout to the driver as clearly as possible. If trees are planted on either side of the least important of the two routes this will help to convey its secondary role since the motorist, if he takes this route, will have a sense of going out of one type of space and entering into another, **8.13**. Careful detailing of the layout to enable right turning movements to be made safely is obviously important here also. In a similar manner vegetation can be used to emphasize the existence of a slip road on the near side of the main route, **8.14**.

If trees are grown on a roundabout this will present the approaching driver with a visible barrier which will induce caution. Planting on either side of the approach road, by creating a sense of enclosure, will also encourage a reduction of speed, **8.15**. The trees on the roundabout in photograph **8.16** constitute an unmistakable barrier.

8.16

8.17

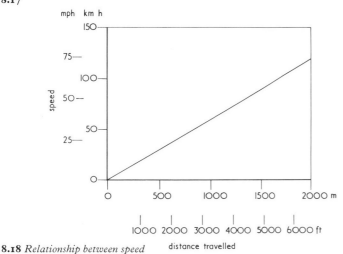

8.18 *Relationship between speed and distance travelled during one minute*

With some alignments the planting of tall trees on and in the vicinity of a roundabout or interchange has the additional advantage that it can be seen for a long distance along the approach roads, thus giving the driver advance warning of the obstacle.

The sense of enclosure created by moving into a zone where the trees grow near to the road edge is illustrated in **8.17**. The inducement to reduce speed here is entirely appropriate, since the motorists are moving from the open countryside into the busy centre of Welwyn Garden City, Hertfordshire. Equally appropriate is the high degree of maintenance evidenced by the well cared for trees and the close-cut grass, since this length of road constitutes a transition zone between the countryside and the town.

It is worth noting that the use of such a transition zone is not invariably the best solution to the problem of effecting the change from one type of environment to another. A sudden transformation from landscape to townscape has great attractions and is to be preferred in many circumstances. It is also more in the European tradition than a gradual change.

As has been noted, a well-designed road will be orchestrated so that the traveller enjoys a varied sequence of spaces and visual experiences. In evolving this sequence the designer will emphasize the best views and suppress the worst ones.

When presenting views to the road user the speed of travel must be taken into account and a reasonable length of time allowed for the view to be appreciated. At 112 km/h (70 mph) a vehicle travels 1867 m (6125 ft) in one minute, whereas at 24 km/h (15 mph) it will travel only 400 m (1312 ft) in the same time. For any particular viewing time there is a linear relationship between speed and length of viewing gap required. For a viewing time of one minute this relationship is presented in graph form in figure **8.18**.

The designer must also bear in mind that, as has been mentioned, the apparent speed of objects in relation to the vehicle increases with the proximity of the object to the road. Thus a building which is not visible as it is approached and which is sited beside the road will flash past if the vehicle is travelling at say 112 km/h, whereas a building on a hill 10 miles away could be kept in view for a considerable time at the same speed.

Much more consideration than is normal could be given to the use of shrub and tree planting on and around the approach embankments to overbridges. This is especially true of accommodation bridges which connect two areas of severed farm land. The approach embankments, which are often steep and ungainly, can be transformed in appearance by careful planting.

For other bridge types the approaches usually have a flat gradient which can mean that the earthworks are seen from a considerable distance on either side. When this presents a boring prospect the view may be enlivened by vegetation and a fitting setting can be given to what are very often handsome structures, **8.19**.

There are many ways in which views can be framed by planting at the roadside. Considering the case where the road is flanked on either side by trees, the simplest way is to leave a gap which is scaled to the design speed, **8.20**.

The profile of the planting could be controlled by using smaller trees and shrubs so as to gradually reveal the view, **8.21**.

Alternatively the spacing of the trees could be increased so that, at the design speed, the view is not seriously interrupted. This could be done using regular spacing, **8.22**, or irregular spacing, **8.23**.

Clearly there are many possible variations on this theme. The most satisfactory is likely to be one which is dictated by the needs of the particular site being considered.

Single rows of trees on the roadside exhibit a dual effect which is sometimes of special use to designers. When viewed along the road the trunks of the trees present a more or less solid wall, depending on their spacing and proximity to the viewer, **8.24**. Yet when they are viewed at right angles to their line they can be arranged to present little visual obstruction, **8.25**. This enables the designer to draw the attention of the traveller to the view ahead and at the same time to allow the sight of secondary views at right angles to the line of motion.

When the rows of trees line a road on a curve the effect

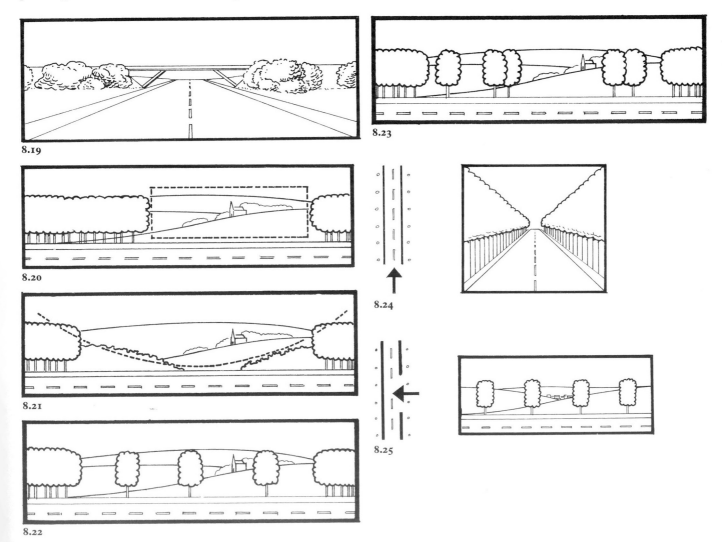

8.19

8.20

8.21

8.22

8.23

8.24

8.25

tends to be the presentation of a solid wall ahead of the observer on the inside of the curve but to give a rather more open prospect on the outside of the curve depending on the radius and the spacing, **8.26**.

Sunlight shining from behind a row of trees at the roadside is likely to be experienced as a flickering at the edge of the driver's vision. Over a short distance and at low speeds this is not important but over a long distance, especially on high speed roads, it can be a major irritant to the driver and is to be avoided.

In the countryside the patterns created by tree growth are the result of the evolution of natural communities or of man's farming and forestry activities. Formal patterns of planting tend to look out of place in such surroundings and so regular spacing of trees in rows would only be used in special circumstances.

8.26

In **8.27** the avenues of trees mark the formal approach to the distant town; by imagining the painting without the trees it can be seen to what extent their presence radically alters the character of the view. The vertical lines of the trees, as well as implying the two planes which contain the road, counterbalance the horizontal lines of the landscape and create a whole which is in equilibrium.

(Reproduced by courtesy of the Trustees, The National Gallery, London)

186

Distant views along the line of the road can be framed in different ways by planting. Three examples are shown in **8.28**, **8.29** and **8.30**. Obviously the provision of a canopy across the road is only possible with relatively narrow roads, and when problems relating to leaf and branch fall and frost pockets can be successfully overcome.

Circumstances, at times, dictate the adoption of an alignment which is not ideal and results in the appearance of a kink or discontinuity from some view-points along the road. If this happens the designer may be able to control the unsatisfactory view by the use of planting so that the short-comings are not visible. Such a case is illustrated in **8.31**.

Plants can be used to screen a road so that it does not dominate views from outside its boundaries. For example when the road runs parallel to the contours and cuts into a hillside, trees positioned as in **8.32** will effectively disguise the presence of the cutting.

Trees and shrubs can also be used effectively to screen an embankment by growing them at the original ground level and on the side slopes, **8.33**.

8.30

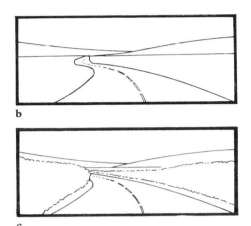

8.31a

b

c

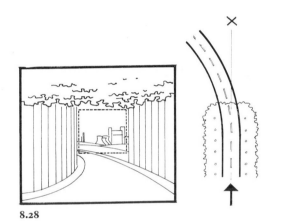

8.28

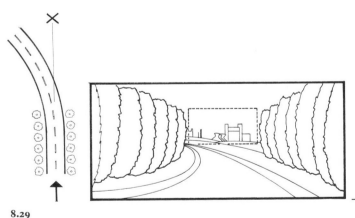

8.29

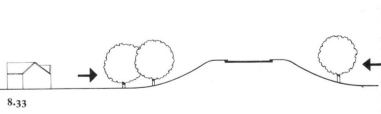

8.32

8.33

8.34

Planting and townscape

ARCADIA

In the Arcadian environment the vegetation and not the buildings are dominant. The road is experienced as passing through a soft landscaped area with the structures glimpsed amongst the foliage. Obviously in such a setting the planting design is of major importance in determining the success of the road as an integral and enjoyable part of the environment. The aim of developments of this kind is usually to afford the dwellers the advantages of town life at the same time as they enjoy the pleasures associated with being in the country. The Arcadian planting in relation to roads then can be informal and imitative of country arrangements of vegetation or it can be semi-formal with a greater emphasis on the architectural use of plants. An example of the former type of solution is that which has been used at The Brow at Runcorn and is illustrated in **1.45** to **1.52**, on pp 25–27, and the latter type of planting is to be found at Welwyn Garden City in Hertfordshire.

Virtually all the architectural methods of controlling and articulating space along a length of road which are referred to in this book can be realized with the use of planting as well as with the use of structures. Since the principles are essentially the same for both media it will not be necessary here to consider planting in detail in this regard, but a brief analysis of **8.34** will serve to illustrate the point.

Some will argue that the high level of maintenance in evidence in this photograph is not feasible today (although it was taken in the summer of 1976) and that therefore the architectural approach to planting design along routes is not possible. This is invalid since, even if the premise regarding maintenance were correct, the same principles can be applied with the use of low maintenance planting by the careful selection of suitable trees and shrubs which require a minimum of attention and the judicious location and configuration of grassed areas to make possible the most efficient use of cutting machines.

In examining this photograph there are a few points worth noting which, although not related directly to the planting, nevertheless enable it to make its full impact, **8.35**. There is a welcome absence of unsightly signs, parking meters and other clutter. The lamp standards are an admirable design and their unassuming simplicity make them an asset in the view. By the application of a light-coloured aggregate to the footpath the designer has avoided the overpowering effect of excessive areas of asphalt. By not using a white precast concrete kerb the devisive over-emphasis of the linear quality of the road has been avoided, so that the whole ground surface can be seen as a series of areas of varied colours and textures.

Suburban areas can take on an Arcadian character if the planting is sufficiently rich and profuse. In **8.36** the foliage of the trees along the road edge merges with that of the front gardens so that the vegetation completely dominates the prospect along the road and the closely spaced dwellings, although not more than 7 m from the footpath, are only visible on transverse views along the route.

The Arcadian impression is strengthened by the use of different species of trees bordering the street, breaking down the formality which results from having the same species in two rows, one on each side.

The trunk and foliage of the chestnut tree together with the mass of foliage on the opposite side of the road define the sides of the space through which the road passes.

size, colour and habit of small tree adds interest and minor focal point.

hedge continues across the road providing a barrier through which road passes. Thus viewed length of route divided into two spaces arranged in series.

Poplars provide major focus and vertical, linear element in pleasing contrast to rounded volumes of foliage, transverse planes of hedges and horizontal linear elements of road and footpath.

Tree species with lobe foliage presents additional linear vertical contrast to foliage volumes and planes.
Interesting contrast between the elegant, machined, line of lamp standard and organic lines of tree.

Transparent screen gives extended glimpse of view beyond.

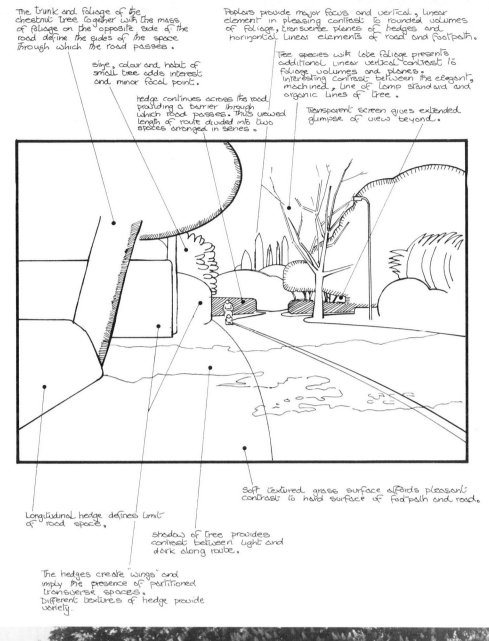

Soft textured grass surface affords pleasant contrast to hard surface of footpath and road.

Longitudinal hedge defines limit of road space.

shadow of tree provides contrast between light and dark along route.

The hedges create "wings" and imply the presence of partitioned transverse spaces.
Different textures of hedge provide variety.

8.35

8.36

LINEAR LANDSCAPES

In urban areas as well as in the very low density Arcadian setting it is sometimes possible to design the road corridor as a linear landscape. By allocating a sufficiently wide strip of land for road construction the road surround can be planted in whatever manner is deemed suitable for the area. This arrangement has the advantages of providing a pleasant environment for the motorist, enabling pedestrian and bicycle routes to be located away from the traffic in park-like surroundings and at the same time distancing the vehicles from the dense urban areas with the accompanying benefits of relief from noise, pollution and the sight of traffic. At the present time the distributor roads for the new towns are often located along landscaped routes which pass between the housing neighbourhoods. Consideration could be given in the future to the development of routes along green corridors in our older cities. Areas which are undergoing slum clearance and comprehensive redevelopment could have such routes incorporated into the new construction. This work could be designed in such a way that the route would eventually be part of a network of greenways covering the whole city area and connecting up the parks, squares, greens and commons already in existence and those new metropolitan landscaped spaces planned for the future. Such green corridors need not be more than say 30m wide on average, so that they would be easily crossed transversely by pedestrians moving from one urban area to another on the other side of the linear parkway. Much narrower widths could be used over some lengths; the width need not be constant, so that a variety of landscaped spaces could be created.

The possibilities in planting such greenways are endless. They can be developed as ecological corridors within the city. Natural plant groupings could be established and allowed to develop, providing an attractive habitat for small wild animals and many species of birds as well as being an exciting foil to the adjacent urban areas; some lengths could be planted as woodland, others in the manner of a landscaped park, and so on. Three ways of planting trees in relation to such a roadway are mentioned below.

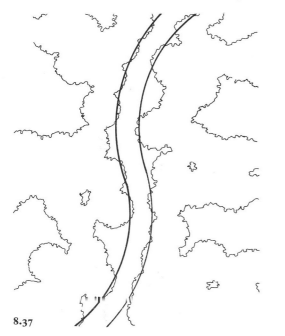

8.37

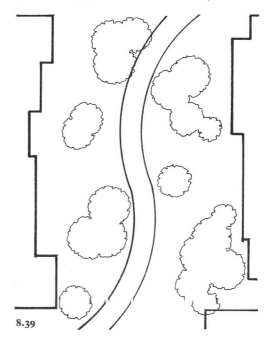

8.39

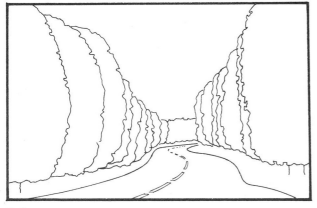

8.38

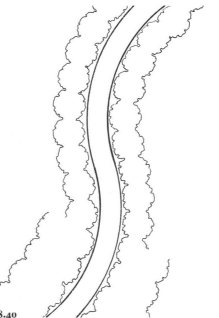

8.40

8.41

8.42

8.43

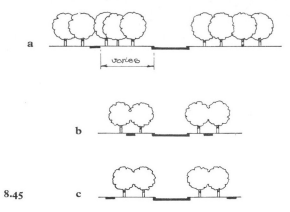

8.45

One way of planting trees in relation to roads, when sufficient area can be made available, is to provide them in such numbers that the impression is made of the road moving through woodland, **8.37**. The trees then contain the space through which the road passes, **8.38**, and at the same time it creates a visual boundary to the spaces on the other side of the tree belt.

Such planting can link up with that in landscaped spaces free of traffic and with that adjacent to other roads, thereby acting as a powerful unifying element for a whole neighbourhood.

Providing a wide planting area on either side of the road has the added advantage of enabling large, long-lived trees to be used in a situation where they will not require more than minimal maintenance once they are established.

A more open impression can be achieved by placing the trees in informal groups with some single trees, **8.39**. When this type of arrangement is used, consideration must be given to the way in which the groups subdivide the overall space and to those views which are opened up and concealed.

A third method is to plant the trees in rows parallel to the road, **8.40**. Single or double rows may be used depending on the space available and the scale of effect required. The best results are often obtained when the trees are spaced close enough for their canopies to meet along the rows and in some cases also across the road. This configuration screens the traffic from those other users of the parkway who are not in vehicles as well as from the urban areas on either side, at the same time leaving relatively large open spaces between the buildings and the road if these are required.

The use of single rows of trees on either side of a road traversing an urban open space is illustrated in **8.41**. The traffic is clearly separated from the open space yet the motorist can enjoy the view. The dramatic contrast in the quality of the light inside and outside the road space adds greatly to the visual pleasures of the area. Seen from within the common the line of the road is transformed from an unwanted visual intrusion to a pleasant wall of vegetation enclosing the communal area, **8.42**. Such an arrangement is clearly adaptable to a linear parkway within a city. By setting the trees further back from the edge of the road the canopy can be prevented from overhanging the running surface. In doing this the scale of the space containing the road is considerably altered, as is illustrated by **8.43**. The use of twin rows of trees is illustrated in **8.44**. In this example the pedestrian as well as the motorist is given the advantage of a separate, shaded, tree-lined route.

The provision of routes for pedestrians and cyclists adjacent to that of the road will often be required. The options open will depend on the width of planting strip or parkway provided. As has been mentioned, when this is substantial it may be possible to provide pedestrian and cycle paths which, although following the same direction as the road, are not kept parallel but, at some distance, wend their way through a landscaped area, **8.45a**.

When two rows of trees are provided on the same side of the road the footpaths can be placed between the rows of trees or on the opposite side to the road rather than on the same side, **8.45b** and **c**.

ROADSIDE TREES IN RESTRICTED SPACES

When the overall width available for roadworks and attendant planting is restricted, a planting pattern comprising single rows of trees is often adopted. This type of layout is applicable to curved roads as well as straight ones. Straight rows of trees are commonly incorporated when the road pattern comprises a rectangular grid network and in the correct circumstances they can greatly improve the appearance of such areas. For example where the road is lined with detached or semi-detached dwellings of different designs the trees can act as a unifying element as well as linking the road to communal open space.

Even with a single row of trees the character of the walkway and the road are greatly dependent on the width of the footpath and the distance of the footpath from the kerb. This can be seen by comparing **8.46** and **8.47**. In the former example the sense of spaciousness resulting from the side verge and footpath area contrasts with the confinement of the pedestrian to a narrow strip between the trees and the front gardens in **8.47**. This is not to say that one solution is better than another, the point is simply being made that what would be an apparently small difference in width on a drawing board results in a large difference in the character of the route.

If the tree spacing is sufficiently close the spatial volume allocated to the pedestrian will be as clearly defined as the space contained by an arcade or a corridor inside a building, **8.47** and **8.48**. Such an arrangement gives the pedestrian a strong sense of refuge from the elements and the traffic. When the longitudinal spacing of the trees is such that their canopies interlock then the pedestrian space will be strongly defined by deciduous trees in winter as well as in summer.

Even in the case of trees which have been severely pruned the effect is quite pronounced, as can be seen in **8.49**. Here the combination of roadside trees with the planting in the gardens transforms the scene. Without these elements the footpath would be lined simply with a high wall on one side and parked cars on the other.

A row of trees need not always line the whole length of the road. In a short street this is preferable but on a long stretch of road the change of scale and sense of containment which is experienced by moving from part of the road without trees into a tree-lined length adds variety and interest, **8.50**. The pleasurable contrast between open and closed spaces can be enjoyed by both the motorist and the pedestrian by starting the row some distance along the road.

This method of planting has the added attraction that the wooded area presents a beckoning goal as it is approached. It is desirable that such a row of trees marks some change in the configuration or usage of the road; for example it could be placed opposite a communal open space, a row of dwellings that has a particularly deep setback from the road or, as in **8.50**, opposite a row of shops and restaurants.

8.46

8.47

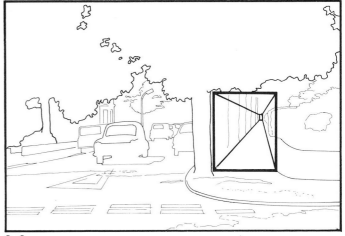

8.48

8.49

8.50

It is sometimes feasible with distributor and even access roads to incorporate a central reserve suitable for supporting a row of large trees. Such an arrangement is only acceptable where maintenance is regular and meticulous and traffic speeds are not high. Fallen leaves should not be allowed to accumulate because of the danger of skidding. Also the crowns must be kept above the tops of the largest vehicles and regular removal of old, diseased and damaged branches must be undertaken. Some form of barrier should be used between the trees and the traffic; initially to protect the trees and later to protect the traffic. In **8.51**, a kerb and low retaining wall perform this function as well as being in character with the surroundings.

Relatively narrow reservations, suitable for high density urban areas, can sometimes be used successfully for this purpose, **8.52**.

The scale of the space between the buildings is quite different in **8.51** and **8.52**, yet the planting and the structures complement each other in both cases.

When a single species is used to line a road the character of the place will be greatly influenced by the choice; compare the effect of the chestnut trees, lime trees and Japanese cherries in **8.54**, **8.55** and **8.53**.

Where columns of trees are used architecturally on each side of a straight length of road they will only look entirely satisfactory if they lead the eye to a focal point or focal area at the end of the road. It is also usually important that their presence defines the sides of a contained space. The other sides of the space can be contained by structures, **8.56b**, planting, **8.56c**, or land form. It is not always possible to have a dominant focal point at the end of a tree lined road and in this case if the containing elements are of adequate scale this is sufficient to achieve a satisfactory effect. In the example of **8.57** the building at the end of the road is not sufficiently substantial to act as a focus or to bring about adequate enclosure.

8.52

8.51

8.53

8.54

8.55

8.56a   b   c

8.57

195

The planting of trees at a regular spacing in rows along the road is by no means the only way they can be used when the overall width of the road corridor is restricted.

One alternative is to locate the trees in small groups with an occasional single tree along the length of the road, 8.58, resulting in a less formal appearance and rendering the space through which the road passes more static in character. In residential areas it may be possible to group the roadside trees together with trees growing in the front gardens.

It is not necessary to use the same species throughout, although in order to avoid a restless, spotty effect, one species should be dominant. In this type of arrangement, as elsewhere, trees with brightly coloured foliage should be used sparingly. They are best employed to provide an occasional contrast to the overall green hue of the roadside foliage or as single specimens creating a focal point in a composition.

Two more methods of incorporating roadside trees in residential areas are indicated in 8.59 and 8.60, reproduced from the Warwickshire County Council publication *Space Relations in Housing Areas*.[27] Sketch 8.59 shows the use of a variable width verge between the road and the front gardens whereas 8.60 illustrates the incorporation of planting beds between hard standings and other hard landscaped areas. Whatever method is adopted, careful positioning of the dwellings in relation to each other, whether detached, semi-detached or terraced, and in relation to equally carefully arranged tree planting can result in a series of interesting and varied spaces along the route.

Another way of incorporating trees is to arrange them in small clumps in spaces created by recesses in the building line, 8.61. This expedient is particularly appropriate in highly urban areas where the buildings are the primary element in the roadscape.

Of all the factors affecting the impact of the tree on its surroundings one of the most important is its height and spacing in relation to the space in which it is placed. Large trees placed close together have a major architectural impact; they enclose space. Small trees placed relatively far apart function in quite a different way. They are experienced as objects within a space, rather than being themselves space-forming. This is a crucial difference which must be taken into account by designers.

An example of the use of trees as incidents in a larger space contained by planting is shown in 8.62. Here even the presence of a row of trees, (on the right hand side), because of their small size, only modifies but does not completely divide the larger volume.

The intermittent planting of single trees also considerably modifies a space contained by buildings yet still enables the observer to experience the overall volume as a single place, 8.63. With this type of usage the apparent modification of the space is likely to be different from different viewpoints. Whether or not the tree becomes a purely decorative element is a function of its size in relation to that of the overall space. An important aspect of this is the height of the tree in relation to that of the general building elevation which defines the 'ceiling' of the contained volume.

A single large tree strategically placed can have a visual impact in a roadscape out of all proportion to its apparent significance on a plan. The holm oak on the left hand side of the road in 8.64 provides a mass which balances that of a long row of trees at the opposite kerb. The relatively small buildings, if they stood alone, would appear out of scale with the canopies opposite. Being evergreen the holm oak is equally effective during the winter months.

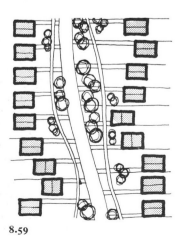

8.58

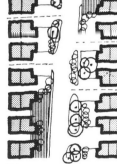

8.60

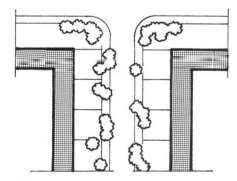

8.59

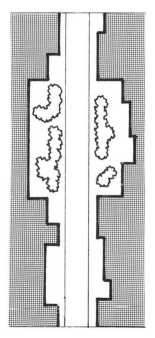

8.61

196

8.62

8.63

8.64

8.65

8.66

TREES AND BUILDINGS

Trees have a major role to play in the integration of views
from the road in which visually unrelated buildings or
groups of buildings present a disjointed effect. One way of
rectifying such a situation is to plant stands of trees in
strategic positions to unify the disparate elements. A bold
approach is often required. Whereas the planting of one or
two isolated trees will only intensify the restless aspect, the
use of tree masses can be a powerful unifying element. An
example of this is shown in **8.65** and **8.66**, where it can be
seen that a considerable improvement would be achieved
by filling the gap between the buildings with a mass of trees
and replacing the foreground bollards with a row of shrubs
or trees. Such expedients would impart a much greater
sense of containment to the road users as well as controlling
the views.

When the open spaces are reserved for other uses an
alternative to block planting is the use of rows of trees to
screen the gap, **8.67a** and **b**.

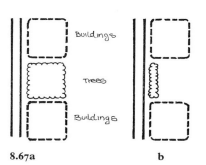

8.67a   b

8.68

8.69

The use of a row of trees to provide continuity along the road edge by screening an open space is illustrated in **8.68**.

Another example of this usage is shown in **8.69**. Here the trees enclose the open space, contain the vólume of the road in a similar manner to the buildings in the middle distance and at the same time frame the distant vista. In **8.69** the trees continue the façade of the buildings facing the street. Effectively they function in the same way at the end of the terraces in **8.70**, preventing the enclosure being dissipated through gaps between the ends of the terraces and the focal building.

8.70

199

Although the trees in **8.70** serve the function of carrying the plane of the facing building elevation over to the terrace blocks, they are, in fact, mainly planted behind the building. From a visual point of view, then, planting which forms a backdrop to a structure can have a similar containing effect to planting at the sides. This principle can sometimes be applied in suburban situations where detached or semi-detached buildings line a road. Tree planting behind the dwellings but visually filling the spaces, either in the back gardens or beyond, can help to unify the views from the road.

In the choice and positioning of trees in relation to buildings the relative size of trees and structures and the distance between them must be carefully considered. The way in which different height/width ratios are experienced is discussed in Part I; similar considerations hold for the spaces bounded by trees or by trees and buildings.

When the height/width ratio for space enclosing structures is too small to be effective the planting of in-scale trees can rectify the matter by sub-dividing the overall space into a number of smaller volumes of more pleasing proportions, **8.71**.

But if the trees are too small they will be ineffective, **8.72**. If trees are to be used decoratively in relation to a road then the other elements of the environment should be strong enough to create spaces within which these trees can be set. Rows of large trees close to the front of small houses are likely to detract from the scale of the dwellings, **8.73**.

With medium rise urban buildings majestic large-scale spaces can be formed by the use of major forest trees, **8.75**. In this example the London plane trees, which have been able to attain their full stature in their situation in a London square, are a perfect foil for the five- and six-storey blocks containing the square.

But it sometimes happens that the buildings have been placed so far apart that a feasible procedure would be to use large trees to form a space with a different scale from that between the buildings and the road, **8.74**.

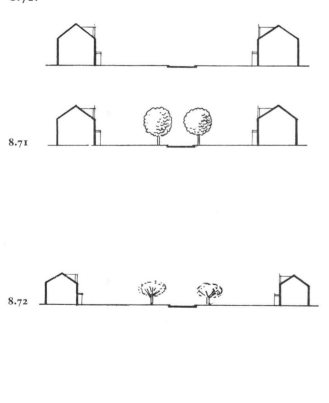

8.71

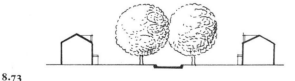

8.72

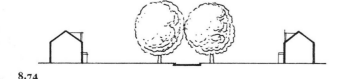

8.73

8.75

8.74

200

Of course, trees need not always be the same height as nearby buildings in order to relate to them. Although a flowering cherry tree is likely to look absurd as a primary landscape element at the base of a twenty storey tower block there are nevertheless ways in which relatively small trees can be used in relation to buildings with satisfactory results.

One factor which should be taken into account when considering the view from the road is that the apparent relative size of objects is a function of their distance from the observer. Thus trees close to the road can appear to be in scale with much larger buildings at a greater distance, **8.76**. In this way trees used in rows or massed close to the road can be effective architecturally in medium and even high rise surroundings.

When the buildings are close to the road small and young trees can relate in terms of contrasts and harmonies of form, texture, volume and so on, **8.77**. In this photograph, which shows one of a row of sycamore trees, the organic lines and substance of the plant compliment the geometric lines and inorganic materials of the buildings.

In **8.77** the large expanses of glass and the distance of the tree from the structure should prevent any problem of overshadowing. But when the windows are less extensive and the tree is close to the building, **8.78**, the use of a more lightly foliaged tree is advisable. This one has been carefully pruned to lighten the canopy and encourage vertical growth.

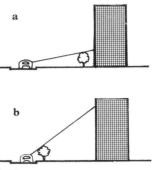

8.76

8.77

8.78

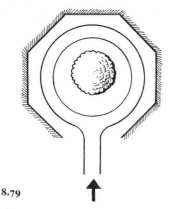

8.79

Single trees can play a variety of roles in relation to the spatial qualities of road alignments. A solitary tree, for example, can often be used as a focal point for a small group of dwellings facing onto an access road, **8.79** and **8.80**. The tree in **8.80** is also the axis on which the loop at the end of the short access road revolves and the visual goal of those approaching the space along the road. An arrangement such as this, providing as it does a large area of grass and topsoil round the tree, enables a forest tree to develop fully since it will be amply supplied with water and nutrients and there is no restraint on root growth or the formation of the canopy.

8.80

8.81

8.82

Another use of a solitary tree is illustrated by **8.81**. Here the purpose of the false acacia is primarily to supply a decorative incident within the space defined by the buildings. It has also the functional attribute of softening the visual impact of cars parked in front of the garages beside the front doors. Single trees are used as decorative elements in **8.82**. The character of this attractive space, contained mainly by the asphaltic materials of the floorscape and the hard vertical planes of the building elevations, would have been significantly harsher without the softening influence of the trees. Once again, the qualities of the organic and inorganic materials are heightened by their juxtaposition. The association of the spherical volumes of the trees with the cubic volumes of the structure also adds interest to the scene.

8.83

TREES AND PARKING

When cars are parked on, or adjacent to, a road the presence of trees will always soften their visual impact. This will be the case whether they are in the shade of trees which cast a dense shadow, like the horse chestnuts in **8.83** or they are overshadowed by much lighter foliage such as that of London plane trees. Most species of lime are unsuitable for use in the vicinity of parked cars since aphids feed on the leaf sap and exude 'honey dew' which will fall on the vehicles. *Tilia euchlora* may be considered for use in such areas since it appears to be shunned by these insects.

Parking spaces placed adjacent to the road and at right angles to it will greatly increase the overall width of hard surface especially when there are large numbers of spaces side by side, **8.84a**. This can cause the asphalt or concrete surface to appear out of scale with the surroundings. A great improvement in appearance can result if strips for shrub and tree planting are placed between small groups of the hard standings, **8.84b**. Sometimes it is possible to incorporate quite long bands of planting between the groups and in residential areas especially, an irregular pattern can be advantageous since it dissipates the look of regimentation which may otherwise result, **8.84c**. The parked vehicles are likely to be even less conspicuous if a serpentine alignment is adopted for the road, **8.84d**.

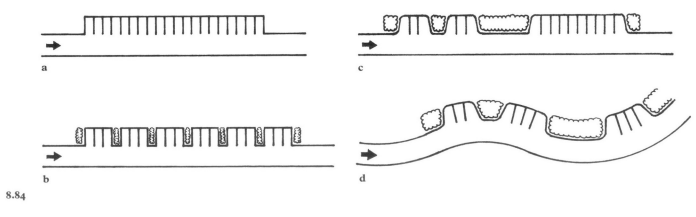

8.84

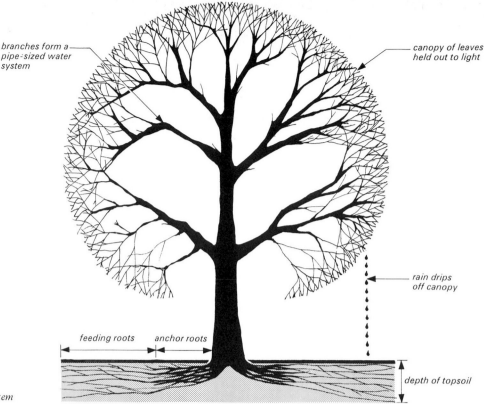

branches form a
pipe-sized water
system

canopy of leaves
held out to light

rain drips
off canopy

feeding roots   anchor roots

depth of topsoil

8.85 *Root and branch system*

TREE CARE

Designers who are aware of the great contribution trees make to townscape will continually seek opportunities to plant new trees in suitable urban settings, particularly large forest trees when the scale of buildings and spaces is sufficiently large.

The choice of species is conditioned by many factors including the location of the trees in relation to the structures and open spaces, the scale of the buildings, and the species already growing in the neighbourhood. Publications which recommend suitable species are readily available and the titles of some of these are included in the select bibliography at the end of the book.

There are various views regarding the most suitable trees for urban sites and the designer is advised to aquaint himself with as much of the written material on the subject as possible. One publication which deals briefly with the topic is the Design Council's *Street Furniture from Design Index*.[28] It says that 'As a guide, small trees are no substitute for large ones in an urban landscape. Forest trees will dominate the most chaotic street scene and create a coherent landscape where the buildings have failed. They provide a background against which smaller, possibly flowering species can be set'. Later it says that 'Best of all are the plane, lime and horse chestnut,' and there is no doubt that these are most important urban trees. Of course, they should not be selected to the exclusion of all the other trees which have been used successfully in various urban settings.

It is most important that the designer is aware of the tree species which already exist in the neighbourhood of proposed new planting. The character of local urban areas owes much to the choice of species which has been made in the past and new planting should continue the existing tradition. An arbitrary choice of trees will fail to integrate the new soft landscaping into the old and furthermore the existing vegetation is a good indicator of those species which will grow most successfully in the area.

There is some controversy about the safe distance of trees from buildings in different soil conditions. Here again the designer should read extensively on the subject and familiarize himself with the behaviour of trees in relation to buildings in the area with which he is concerned. When making a decision it should be borne in mind that normally the only cases in which difficulties may arise will involve planting in shrinkable clay soils. (There seems to be a tendency for recommendations made for planting in such soil to be applied to all types.) If trees which take up a large quantity of water (particularly willows and poplars) are planted close to buildings on clay soils they can cause a marked reduction in the water content and thus reduce the volume of the soil. This may result in differential settlement of the foundations and consequently cause or contribute to cracking in the superstructure. There is a deficiency of research in this field which is now being rectified.

Many of the roadside forest trees in urban situations grew to maturity before the use of tarmac or concrete for road pavements. They therefore grew up in much more favourable conditions than trees planted today in areas with an impermeable surface.

Trees have a radial root system which does not usually penetrate more than 600 to 900mm below the ground surface. Some also have vertical roots to assist in anchoring the tree. The portion of the radial roots closest to the trunk are also anchor roots, while the water-collecting portions, called the feeder roots, occur towards the end of the radial system below the edge of the canopy, **8.85**. The feeder roots should be in top soil; water must be able to penetrate the soil and sustain the soil bacteria, especially those engaged in the fixation of nitrogen. If waterproof surfacing comes up to the trunk of the tree it will not receive the

8.86

nourishment it needs. In the planting of new trees, therefore, as large an area as possible of pervious surface should be left round the trunk. It is only in locations where this area can be quite extensive that forest trees can be expected to reach their full stature. In relation to roads such locations could include wide verges, wide central reservations, landscaped squares and so on. Nevertheless it is often worth planting such trees in conditions which are less than ideal where, with proper care, they will attain a more limited but still substantial size.

The well-being of trees, especially in the vicinity of roads, requires the attention of competent experts possessing a high degree of specialized knowledge and practical skill. This attention will include the removal of dead and diseased branches, crown lifting, crown thinning and the reducing and reshaping of the crown. The monstrous appearance of the trees in **8.86** is totally avoidable by the proper execution of these operations. This form of tree mutilation is called lopping. The tree never recovers a natural form after being treated in this way.

BRANCH REMOVAL In order to avoid damage to the tree, branches should not be removed with one cut. If one cut is used the branch is likely to tear off a strip of bark as it falls. Also the branch should be removed in such a way as not to leave a stump. This would not heal over and would allow access to water, insects and fungal spores, causing the wood to rot. Branches are properly removed using three cuts as illustrated in **8.87**.

CROWN LIFTING Trees situated close to roads may need to have their crown lifted in order to prevent overhanging branches obstructing vehicles. Other reasons for lifting the crown include the wish to open up a view or let in light under the canopy either to cut down overshadowing of buildings or of shrub and groundcover planting below the trees. The operation entails removing the lower branches back to the trunk in the manner outlined above. When this operation is properly carried out the tree will continue to be well proportioned, **8.88**.

CROWN THINNING When trees are situated close to buildings this operation can result in considerably more light penetration. Another reason for carrying out this procedure is the desire to reduce the wind resistance and so reduce the likelihood of damage to the tree in strong winds. The operation entails the removing of some of the small branches back to the trunk as well as the removal of parts of branches, **8.89**. The aim should be to maintain the same shape of head and an equal distribution of branches throughout.

CROWN REDUCING Reducing the head may be carried out to lessen the obstruction of light or of a view, to prevent the tree approaching too close to overhead cables or a building, or simply to improve the proportions of the tree in relation to its surroundings. Reducing involves the removal of parts of branches as in **8.90**. The objective should be to lessen the overall size of the crown at the same time maintaining a natural, well-balanced shape.

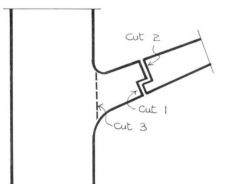

8.87 *Branch cutting : Cut 1— undercut ; Cut 2—main cut ; Cut 3—stump cut off flush with trunk*

8.89 *Crown thinning*　　Before　　After

Additional reduction to allow for overhead cables, etc.

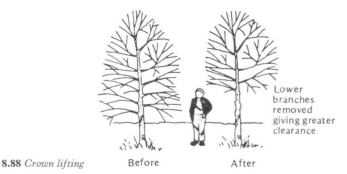

Lower branches removed giving greater clearance

8.88 *Crown lifting*　　Before　　After

8.90 *Reducing head*　　Before　　After

8.91

SHRUBS

Although trees are by far the most important plant in relation to roads, shrubs also have many vital roles to play. Some of these have already been mentioned: the formation of ecological communities of plants at the woodland edge of roads in the countryside; the planting of hedgerows instead of building fences when the use of hedges is the local way of containing land; as a screen between pedestrians and traffic. There are many other ways in which shrubs can be used and some of them are explored here especially in relation to residential developments.

One way in which shrub planting can contribute to the character of a residential area as well as fulfilling important architectural and strategic functions is illustrated in **8.91** and **8.92**, of the Brow at Runcorn in Cheshire. Indigenous species have been planted so that the character of the surrounding landscape continues through the development, as can be seen by the relationship between the background and foreground in these two photographs. The generous use of shrub planting close to the buildings anchors the structures to the land forms and imparts an established air to developments which were completed only a few years ago.

8.92

206

An objection put forward to the use of narrow roads in residential areas is that cars will be parked in the roads by people wishing to use the quickest access to their houses. This, it is said, will result in blockage, long delays and frustrated drivers mounting verges to find a way past. In these schemes the planting has been designed in conjunction with careful ground contouring to ensure that motorists are not tempted to find short cuts in this way. At the same time adequate parking space is available in the vicinity of the fronts of dwellings. The entrance to such a space is shown in **8.93**. Apart from the barrier created by the dense planting, the device of bringing it to the edge of the running surface further discourages parking due to the difficulty of opening the nearside doors. Although the general appearance of the planting is informal it is carefully arranged to delineate and contain the spatial volume allocated to the roads.

The screening of roads from the dwellings is often desirable in housing areas. An effectual and visually pleasing way of doing this is by the use of shrubs. A fine example is shown in **8.94**. This view proclaims the influence of a landscape designer capable of integrating all aspects of the surroundings into a unified whole which is both pleasing and highly functional. The mass of shrubs screen the back of the dwellings from the view of the traffic and the extent of the planting distances the building from the road sufficiently to achieve a significant noise reduction. The contouring of the ground on the opposite side of the road contains that side and forms a mound which balances the volume of the planting. By bringing the planting to the road edge and sloping the ground to the kerb on the opposite side a visual logic has been established for the serpentine road alignment. When the trees on the sloping ground mature they will balance the volume of the building opposite and increase the enclosure of the road. One tree has been placed in front of the gable end thus relieving the view from the road of the large area of brickwork. A coarse grass mixture has been sown giving a natural appearance and cutting down the maintenance required.

Shrub planting can also be used in association with mounding to form a barrier protecting the dwellings from the views and noise of the road, **8.95**. This arrangement has the added advantages that it is a more effective noise shield and a more effective barrier between pedestrians and traffic than shrubs alone.

In residential areas as elsewhere roundabouts often result in large unstructured spaces. Their appearance can be greatly improved by shrub planting. When these are used boldly a sense of spatial containment is achieved and at night drivers are protected from the glare of headlamps. To achieve satisfactory visual results there should be containing elements round the outside of the road as well. Mounding the ground or raising the general level, as in **8.96**, makes the shrubs more effective. The use of bedding out plants or small single shrubs is rarely successful in such locations. They invariably look out of scale as embellishments and they have no space controlling qualities.

8.93

8.95

8.94

8.96

FRONT GARDENS

In residential areas where the front entrances of the dwellings face the road the presence and type of front gardens is likely to have a considerable influence on the character of the road. Front gardens, like the fronts of houses, are usually better cared for than the backs. The front projects the public image of the household. Thus the provision of front gardens can be a powerful vehicle for providing visual interest and variety to a road. This is an area in which designers can learn from the successful solutions of the past as well as the present. Obviously front gardens are more likely to be included in low density than in high density developments and the lower the density the more impact they are likely to have on the environment from the point of view of the road user. There are exceptions to this density correlation; for example in village streets gardens are often not provided and if they are the size is usually minimal. A brief look at some examples will give an indication of the range of solutions which have been used.

When the road runs parallel to a high wall or fence the use of climbing plants can relieve the monotony of the view. This situation often arises when roads are located parallel to back gardens.

In the example of **8.97** the interesting elevations and excellent detailing are supplemented by this usage. The backcloth of large trees enhances the elevations of the buildings; this is made possible here by the grouping of the dwellings around communal open spaces sufficiently large to accommodate trees of this size.

When the shrub height is above eye level the shrubs contain the space occupied by the observer; when the height is below eye level the way in which they modify space is quite different. The space containing the shrubs is experienced in its totality, yet visually their presence affects the space in a number of ways, from the creation of patterns and textures on the floorscape in the case of very low shrubs to the concealment of vehicles and the creation of sub-spaces below the viewpoint in the case of higher ones.

8.97

208

8.98

8.99

In **8.98** the shrubs relieve the
monotony of the flat ground
surface and help to delineate the
front garden and the hard
standing in front of the garages.
In contrast to the effects of
planting at the Brow (**8.91–93**) the
main space between the buildings
is unaltered in size and
proportion. In this case however
the space will be radically altered
in summertime when the trees
have become more mature.
Low planting is often used on
verges and side slopes in
conjunction with grass. As with
other combinations of plants this
can be done in a frankly
decorative or artificial manner or
the designer may, through study
of plant groupings in the wild,
adopt more natural looking
arrangements. In **8.99** the shrub
group has been used to add
variety to the combination of trees
and grass on the verge. Since such
an arrangement of intermittent
groups is mostly viewed along the
length of the road the visual
impact is out of all proportion to
the relative areas of grass and
shrub.

209

Sometimes low shrub or ground cover plants are used instead of grass on verges or side slopes. A planting scheme of this type has the advantage that weeds do not become a nuisance and if the shrubs are carefully chosen almost no maintenance is required, in contrast to the frequent cutting of grass which would otherwise be necessary. It should be appreciated however that the shrubs will need to be replaced sooner or later. Some species may need replacing in as little as five to seven years after planting.

Low shrubs used in this way can also serve the purpose of acting as a deterrent, channelling pedestrian movement in relation to the running surface. They will probably have to be protected with temporary fencing until they are sufficiently mature to discourage movement through them, especially when they cross a desire line. Plants with a strong dense growth should be used for this purpose, **8.100**.

Low planting is often used to assist in the designation of ground areas for specific purposes. In particular it can be used to indicate the areas at the side of a road which should be used for parking. Small beds of plants are frequently sufficient, **8.101**. Such a usage can do much to relieve the appearance of large areas of asphalt or concrete.

If there is danger of the plants being damaged by vehicles or pedestrians they can be placed in raised beds. In some locations raising the beds the height of one brick course could be all that is required; in others it may be desirable to use higher beds so that the plants and planting unit form a screen. In **8.102** this arrangement has been used to screen parked vehicles from the fronts of the dwellings and to create a protected pedestrian precinct between the planting beds and the buildings.

8.100

8.101

8.102

8.103

The small garden in **8.103**, as well as enhancing the appearance of the surroundings, provides a transition zone between the front door of the cottage and the dual use, winding village road. An attraction of this solution is that it enables an environment with a strong visual unity to be formed. It is clearly necessary that vehicles can be constrained to move slowly in such surroundings. A contemporary design might have service vehicle access only to such an area.

In the urban village setting, small gardens provide a semi-private front area and increase the privacy of the dwellings by distancing the windows and front entrance from the road and pavement, **8.104**.

8.104

The larger mature front gardens
of a suburban area can provide an
endlessly varied and interesting
view from the road. In the type of
development of **8.105** the
repetition of the well-articulated
elevation of the dwellings
provides a unifying background
to the varied treatment of the
gardens.

8.105

8.106

tree is responsibility
of householder

minimum problem from
leaf fall on road

maximum privacy
for dwelling occupiers

regular maintenance
of maximum importance

maximum sense of enclosure for users of road

pedestrian distanced from traffic

Trees situated between the front of dwellings and the road
can be placed in a number of locations in relation to front
gardens and pedestrians. Three of the options are indicated
in **8.106**.

213

There is a tendency with infill developments of flats in suburban roads of the type shown in **8.107** to replace the individual front garden with communal space which is open to the road. This must be done with care since it greatly alters the character of the road. In these circumstances it may be possible to maintain the character and proportions of the area by the use of well designed planting and walling. The use of rows of trees can be effective in maintaining the sense of enclosure in the road.

When small private areas in front of the house have to accommodate hard standings for cars their function of enhancing the view from the road is more difficult to incorporate. This is particularly so with narrow frontage dwellings, **8.108**. In this example the trees improve the scene and the use of grass/concrete paving enables the cars to be parked on a grass surface.

8.107

8.108

09

In less dense suburban areas
where large front gardens are
surrounded by walls, high fences
or hedges to increase privacy, the
view from the road is greatly
improved by the use of trees in
the garden and by the roadside,
8.109.

In the Arcadian setting well-
structured spaces can be
maintained even with large open-
fronted gardens, **8.110**, by well-
designed planting; in this case the
use of hedging and carefully
positioned trees has achieved the
desired result.

8.110

Minimal planting

The absence of planting can be as important in establishing the character of a length of road as its use. In village streets, for example, the sense of refuge from the all-pervading landscape outside can be heightened by the exclusion of vegetation. In the tiny village of Inverary in Argyllshire, Scotland, the main street has a highly urban character for this reason, **8.111**.

It is not only in the village setting that restraint in planting is required. Photograph **8.112** shows a view of a housing neighbourhood in Cwmbran New Town, Gwent. Any shrub or tree planting, to be successful, would have to avoid disguising the dramatic forms of the buildings and the bold effect of the repetition of simple shapes which pervades the development.

8.111

8.112

PART FOUR

MATERIALS, SURFACE AND TRIM

9
ROAD PAVEMENT MATERIALS

The road pavement

The treatment of the road surface and of the visible elements associated with the surface are of great importance in achieving a visually integrated environment. The road surface constitutes part of the horizontal plane which is the floor of the outside space. It requires to be designed with the same care and sensitivity as the walls of the urban spaces formed by the elevations of the containing structures. One of the factors which determines whether the road is to be an enhancing or a disruptive visual element is the texture of the surface; another is its colour. In both of these today's designer enjoys a wide choice, enabling him to select the most appropriate material for the character of the place, whether it be sand-paper-textured black rolled asphalt, red brick pavers or interlocking concrete blocks. Modern construction techniques make possible subtle design decisions which can make a significant contribution towards creating a road which blends with its surroundings. For example, instead of using a simple hot rolled asphalt wearing course the rolling in of a particular size and colour of coated aggregate may ensure that the road gradually blends more with its surroundings as it ages and the thin coating of binder is worn off the surface.

Some knowledge of the elements of the total road structure is useful if not indispensible to a proper understanding of the choices available. The appearance of the surface is to a large extent dictated by the materials underneath. This chapter, then, which is concerned primarily with the appearance of the road surface in the townscape alignment, describes briefly the layers below as well as those at the surface. The engineering design of a road carrying vehicles is outside the scope of this book and should be carried out by a properly qualified person.

The most important reference documents on the subject are the following: the Department of Transport's *Specification for Road and Bridge Works*[30] with the accompanying *Notes for guidance on the specification for road and bridge works* and the Department of Transport's *A Guide to the Structural Design of Pavements for New Roads*[31] There is also a *Supplement* to these publications (see p 302 of the Select Bibliography for details) and finally the Department of Transport's Technical Memorandum No H6/78, *Road Pavement Design*, supercedes parts of the four previous references. Together they form the definitive guide to the structural design of road pavements and in effect codify existing accepted good practice.

A consideration of the road pavement will be easier if clear definitions of some of the more commonly used terms in road construction are to hand. These are particularly necessary in the not infrequent cases where the terms used popularly and professionally do not have exactly the same meaning.

After the glossary of terms the basic types of road construction will be described, followed by an outline of the range of surfaces available.

Glossary of terms

Definitions are taken from BS 892:1967, Glossary of highway engineering terms, except where noted. (Material from BS 892 is reproduced by permission of the British Standards Institution, 2 Park Street, London W1A 2BS from whom complete copies can be obtained.)

Aggregate
An inert material such as crushed rock, slag, gravel, sand or the like which forms the greater part of the mixture. Aggregates are described as 'coarse' or 'fine' the dividing sieve size varying for different purposes (see relevant British Standards).

Asphalt
A general term for certain mixtures of asphaltic cement and mineral matter.

Asphaltic cement
Bitumen, a mixture of lake asphalt and bitumen, or lake asphalt and flux oils or pitch or bitumen, having cementing qualities suitable for the manufacture of asphalt pavements.

Asphaltic concrete
Term applied to a particular form of rolled asphalt consisting of a well-graded mixture of coarse aggregate, fine aggregate and, in most cases, mineral filler mixed with a penetration grade bitumen. (Definition from *Asphaltic Road Material* by Hatherly and Leaver.)[32]

Binder
A material used for the purpose of holding solid particles together as a coherent mass.

Bitumen
A viscous liquid, or a solid, consisting essentially of hydrocarbons and their derivatives, which is soluble in carbon disulphide; it is substantially non-volatile and softens gradually when heated. It is black or brown in colour and possesses waterproofing and adhesive properties. It is obtained by refinery processes from petroleum, and is also found as a natural deposit or as a component of naturally occurring asphalt, in which it is associated with mineral matter.

Bitumen emulsion
An emulsion in which, with the aid of suitable emulsifying agents, bitumen is dispersed in water, or in an aqueous solution.

Bitumen macadam
Coated macadam in which the binder is wholly or substantially bitumen.

Bituminous
Containing road tar, bitumen, pitch or mixtures thereof.

Blinding
The application of fine material to a surface, to reduce surface voids or to cover a bituminous binder.

Brick paving
A surfacing composed of bricks laid in a regular arrangement on a prepared road base.

Burnt colliery shale ('blaes' in Scotland)
Shale which has been altered by combustion, usually spontaneous, in a colliery tip or bing.

Cement-bound macadam
A form of road construction in which a mortar of Portland cement and sand is added to coarse aggregate.

Chippings
Single size aggregate of nominal sizes between 3mm and 25mm inclusive. (See also BS 63 and BS 1984.)*

Clinker
Well burnt furnace residues which have been fused into lumps. The term excludes residues from furnaces fired with pulverized fuel.

Coated chippings
Chippings lightly coated with tar or bitumen for use in surface dressing or for rolling into a wearing course.

Coated grit
Grit which has been coated thinly with tar or bitumen binder of such proportions and of such properties as to allow it to be scattered for blinding a wearing course.

Coated macadam
A road material consisting of graded aggregate that has been coated with a tar or bitumen, or a mixture of the two and in which the intimate interlocking of the aggregate particles is a major factor in the strength of the compacted roadbase or surfacing.

Cobble
Rounded or sub-angular stones of sizes between 200mm and 60mm.

Cold asphalt
A close-textured type of coated macadam wearing course material, consisting of aggregate wholly passing 6mm BS sieve for the fine grade, and wholly or substantially passing 10mm BS sieve for the coarse grade, coated with a binder solely or substantially of bitumen, the composition of the mixture being so adjusted that the material can be spread and compacted while cold or warm and, if required, after storage.

---

*References to relevant British Standards and Codes of Practice appear in the Select Bibliography on p. 303.

Cut-back bitumen
Petroleum bitumen whose viscosity has been reduced by the addition of a suitable volatile diluent.

Dense tar surfacing
A hot process wearing course material consisting of aggregate, filler and road tar, in such gradings and proportions that when spread and compacted it provides a close textured impervious mixture.

Emulsion
A relatively stable dispersion of a liquid minutely sub-divided in another liquid, in which it is not soluble.

Expansion joint
A permanent joint between two slabs or elements of a structure which allows a small relative movement perpendicular to the joint.

Filler
A finely-divided mineral powder added to road tar, bitumen or the like, or to a mixture containing the same, in order to effect some desired change in the properties of the binding material.

Gravel
A non-cohesive granular material resulting from the natural disintegration of rocks and consisting of irregular, rounded or angular particles, generally between 76mm and 4.8mm in size. It may contain crushed particles. (See 8228 in BS 892.)

Grit (gritting material)
Small hard fragments of mineral aggregate for application to a road surface.

Gritting
The operation of spreading grit or similar material on a road surface to reduce the possibility of skidding.

Hardcore (rubble)
Recovered stone, broken brick or the like which, when spread and compacted, can be used as a rough filling material.

Hoggin
A naturally-occurring mixture consisting mainly of gravel and sand and containing sufficient clay to bind the mixture together when rolled.

Joint filler
A strip of compressible material used to fill an expansion joint.

Joint sealing compound
A material used to seal the exposed surfaces of joints for the purpose of preventing the entry of water, grit, etc.

Mastic asphalt
A type of asphalt composed of suitably graded mineral matter and asphaltic cement in such proportions as to form a coherent, voidless, impermeable mass, solid or semi-solid under normal temperature conditions, but sufficiently fluid when brought to a suitable temperature to be spread by means of a float.

Mastic asphalt surfacing
A wearing course formed by spreading hot mastic asphalt, with the addition of chippings, by means of hand floats or by machine.

Pulverized fuel ash (fly ash)
Residue from the high temperature burning of pulverized coal in power stations; it is an impalpable dust which may have pozzolanic properties.

Quarry fines
Crushed aggregate passing the smallest screen aperture.

Regulating course (shaping course)
A course of variable thickness applied to a road surface to adjust the shape preparatory to resurfacing.

Road tar
Tar conforming to a specification defining its properties for use in road construction. (See BS 76.)

Rolled asphalt (hot process)
A material used as a dense wearing course, basecourse or roadbase material. It consists of a mixture of aggregate and asphaltic cement which, when hot, can be spread and compacted. (Definition from BS 594:1973.)

Sealing coat
A treatment intended to prevent the ingress of water

Sett paving ('causeway' in Scotland)
A surfacing consisting of rectangular blocks of stone (setts) laid in a regular arrangement upon a prepared roadbase.

Shale
A more or less hard, laminated argillaceous rock, sometimes containing material of organic origin.

Single size aggregate
Aggregate containing a major proportion of particles of one size. (See BS 63 and BS 1984.)

Slag

The stony product resulting from metallurgical processes. The origin of the slag should be indicated by a prefix, eg, blastfurnace, steel, etc.

Surface dressing

A wearing surface consisting of a layer of chippings, lightly rolled into a film of road tar, bitumen or the like, freshly applied to a road surface.

Tack coat

A thin film of binder to improve the adhesion between two courses.

Tar

A viscous liquid, black in colour, having adhesive properties, obtained by the destructive distillation of coal, wood, shale, etc. Where no specific source is stated it is implied that the tar is obtained from coal.

Tar emulsion

An emulsion in which, with the aid of suitable emulsifying agents, tar is dispersed in water, or in an aqueous solution.

Tarmacadam

Coated macadam in which the binder is wholly or substantially tar.

Tar paving

A surfacing of tarmacadam laid in one or two courses for footpaths, playgrounds, and similar areas for pedestrian or very light vehicular traffic.

Waterbound macadam

A form of road construction consisting of broken stone, slag or gravel, intimately interlocked and compacted in the presence of water, the binding agent used being loam, sand, stone particles, or the like.

Wet mix macadam

A roadbase material consisting of graded crushed rock or slag usually pre-mixed with a controlled amount of water sufficient for adequate compaction.

Note: See also the definitions listed under 'The Layers of Construction' below.

Basic road types: flexible and rigid pavements

Road pavements can be of two basic types:

Flexible pavements which have a comparatively low rigidity and depend for their stability and load spreading property on the interlocking of constituent particles such as crushed rock, gravel, and hardcore, and

Rigid pavements which distribute the loads through a relatively stiff concrete slab.

The difference between flexible and rigid pavements is crucial to road design.

In relation to the surface a third category can be postulated: that of *unit paving*. In this type precast units form the top layer of the road construction. Unit paving can be used as the top layer to a flexible or rigid base.

In practice rigid or flexible pavements are used depending on the limiting factors of each particular design problem. A good subsoil may indicate an economical flexible paving whereas a weak underlying ground may indicate the use of a rigid construction. Other engineering factors will include the expected frequency and magnitude of loading together with the design life of the pavement. Economic considerations will include the relative cost and availability of materials and plant. The cost of maintenance must also be taken into account. However in examining all these factors the designer should never lose sight of the general amenity value of the road as an important visible element in the environment. In making the choice of road type, therefore, it should always be borne in mind that the type selected will determine the choice of surface finishes which can be used and the suitability of these alternatives in the context of their surroundings should be assessed at an early stage.

The presence or otherwise of services under the road may also be an important factor. If they are present, ease of access must be considered and also possible difficulties in repairing the surfacing in such a way that the new portions match the old.

Any road construction must fulfil three primary engineering functions. It must:

1 Distribute the loads from the vehicle wheels so that the ground pressure is sufficiently low to ensure acceptably small deflections which are not permanent

2 Provide a satisfactory running surface for vehicles

3 Protect the underlying ground from water and frost, since these may cause deterioration.

Flexible and rigid pavements are the two basic methods by which these objectives are achieved so it is worth considering briefly the different ways in which they act.

If the road construction consists of granular material the particles of which are not bound together (for example crushed rock, gravel or hardcore) then the distribution of the wheel loads will be small compared to that for a highly rigid construction. In an unbound pavement the load spreading properties are dependent on the juxtaposition of particles, the effects of interlocking and the absence of movement between the particles. Each particle is in contact with the ones below and the load causes compressive forces which are transmitted directly from the particles above to those underneath with some lateral spread of load through direct compression, **9.1**. In this respect bound flexible pavements behave in a similar manner to those which are unbound.

On the other hand if the granular elements are firmly bonded together (by using cement, water and a well-graded aggregate to form a concrete slab) the load is distributed over a much greater area. It is characteristic of this type of pavement that it can sustain tensile and bending stresses whereas unbound pavements cannot, **9.2**.

A simplified view of the effects of load on three consecutive hypothetical elements of the two types of road construction may help in an understanding of the difference in load distributing properties.

When a section of the flexible pavement has a load, P, applied there are shear forces, S, set up at the interface with the adjacent material and if these could be sustained internal bending moments, Fx, would be induced. However, even when there is a bonding agent present the shear and tensile stresses induced by bending are too high for the material, which distorts, and the section under load tends to be displaced in relation to its neighbours, **9.3**. In practice, of course, the distortion is slight, extends over an appreciable length of the pavement and the materials revert to their original position after the load is removed.

In the case of the rigid pavement the loads can be distributed outside the area directly below them by means of shear forces and bending movements, **9.4**. The deflected form is more gradual than with the flexible pavement.

Flexible pavements can be constructed of bound or unbound material. The most common bound materials use either a bituminous or cement binder. Those with a cement binder include soil-cement, cement-bound granular base and lean concrete. There are numerous bituminous-bound materials, the main categories being macadams and asphalts. These are described in subsequent sections as are

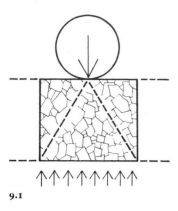

9.1

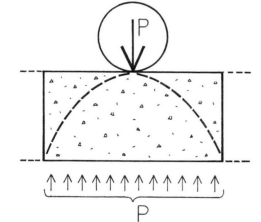

9.2

9.1 *Spread of load through flexible pavement*
9.2 *Spread of load through rigid pavement*
9.3 *Simplified diagram of the effects of load on a flexible pavement*
9.4 *Simplified diagram of the effects of load on a rigid pavement*
9.5 *Relationship of some major road building materials to type of construction*
9.6 *Composition of flexible and rigid pavements*

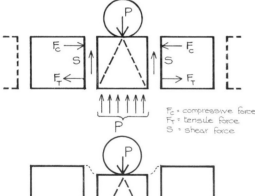

9.3

F_C = compressive force
F_T = tensile force
S = shear force

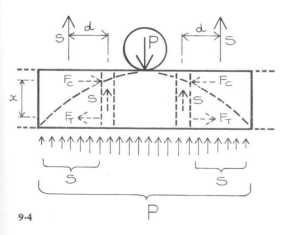

9.4

the other bound and unbound materials.

Diagram **9.5** shows the relationship of some of the most important road building materials to the type of construction. Those bound materials, such as bitumen macadam and lean concrete, which have low flexural strengths are regarded as flexible for design purposes.

The layers of construction

Both flexible and rigid pavements are usually made up of a number of layers. These may comprise all or some of those illustrated in 9.6.

The various layers are defined as follows:

Pavement   The part of the road structure above the sub-grade.

Surface dressing   A wearing surface consisting of a layer of chippings, lightly rolled into a film of road tar, bitumen or the like, freshly applied to a road surface.

Wearing course   The part of the surfacing which directly supports the traffic.

Basecourse   A course forming part of the surfacing immediately below the wearing course.

Roadbase   One or more layers of material constituting the main structural element of the pavement.

Sub-base   One or more layers of material situated between the roadbase and the sub-grade.

Slab   Plain or reinforced concrete which may itself serve as the running surface and roadbase, or solely as a roadbase.

Formation level   The ground level on which the road pavement is constructed (ie, the surface of the sub-grade).

Sub-grade   The upper part of the soil, natural or constructed, which supports the loads transmitted by the overlying pavement.

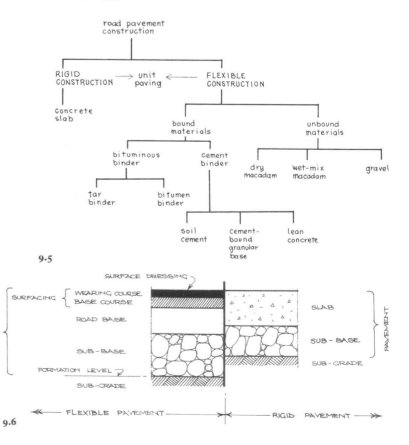

9.5

9.6

It should be noted that a capping layer may be required between the sub-grade and the sub-base. This layer is considered as part of the earthworks and is described in Technical Memorandum No H6/78, referred to above.

There are a number of advantages in constructing the pavement in successive layers.

Since the purpose of the road construction is to distribute the wheel loads it is clear that the top layer will have to sustain the highest stresses and the magnitude of the stresses will diminish progressively towards the bottom of the construction.

The most highly stressed layers are made up with the most expensive materials. It is therefore advantageous to construct in successive layers using, in each layer, that material which is suited to coping economically with the magnitude of stress encountered.

Also, as is clear from the definitions, each layer must fulfil a different function thus calling for a different material. For example a non-skid, small-sized aggregate suitable for a wearing course is unlikely to be the appropriate choice for a roadbase where a larger size material would be adequate and less expensive.

Materials of construction common to basic road types

Fine aggregate and coarse aggregate are major constituents of both concrete and asphaltic pavements.

COARSE AGGREGATE

In the British Isles there is a great variety of aggregates available. The materials used are invariably natural rock and, in the case of asphaltic pavements, slags derived from metallurgical processes. The natural rock material is either that found in massive outcrops or in gravel deposits.

The main sources of aggregate for roadmaking are listed in *Sources of Road Aggregate in Great Britain,*[33] prepared by the Transport and Road Research Laboratory and the Geological Survey and Museum. For each of the major quarries, gravel pits and slag works this publication supplies the name and address of the owner or operator, the name and the situation of the quarry, pit and slag works, the type of stone worked, its group classification and colour.

Sources of coloured aggregates are presented as Road Note 25 in *Sources of White and Coloured Aggregates in Great Britain,*[34] also prepared by the TRRL. The information in this publication is presented in tabular form, and Table 1 lists the best naturally coloured roadstones in production at the time the booklet was published (1966). Samples are contained in the roadstone collection at the Transport and Road Research Laboratory. This collection includes samples from most of the larger quarries producing roadstone in Great Britain.

The roadstones have been divided into six colour groups: blue, green, red, yellow, dark grey and light coloured. Each entry gives details of the colour, the rock type, the county, the TRRL roadstone collection reference, the name of the quarry and the name and address of the owner. Many of the colours are muted since the more brightly coloured rocks are generally of decomposed material and therefore are unsuitable for aggregate. Tables 2 and 3 list white and near white aggregates, including materials that are produced in relatively small quantities.

The definition of stone types is laid down in BS 812:1951, 'Methods for sampling and testing of mineral aggregates, sand and fillers'. It is explained that the rocks are grouped together according to their petrological characteristics and that the group classification is adopted for the convenience of producers and users of stone. These classifications are as follows:

Basalt	Granite	Limestone	Schist
Flint	Gritstone	Porphyry	Artificial
Gabbro	Hornfels	Quartzite	

A classification of natural rocks under these group headings is also given in this British Standard together with a glossary of definitions of rock names.

The Artificial group consists of those materials which occur as by-products of manufacturing processes. They include slags, clinker and burnt shale. Blast furnace slag, the most widely used of the slags for road making, has a texture which varies from glassy to honeycombed.

Gravel is formed by the breaking up of the parent rock by weathering, the further fragmentation by subsequent weathering, and then the smoothing of the particles under the action of flowing water. Consequently many gravels are made up of a large number of different rocks. There are also those which are composed almost entirely of the same type of rock, such as the flint gravels of south-east England and the quartzite gravels of the Midlands. Gravels are characterized by the rounded or irregular shape of the individual stones or pebbles and by the smooth surface and varied size of the particles. For road making purposes the larger stones are often passed through a crusher in order to provide a proportion of angular material.

Coarse aggregate is material retained by a 2.36mm BS sieve; the smaller material passing is classified as fine aggregate.

FINE AGGREGATE

This is either naturally occurring sand or the fine material produced in the rock quarrying process. In sand the most common mineral is quartz and the grain size can vary from 2mm to dust.

To be suitable for road building purposes any aggregate should comply with a number of engineering criteria, depending on the function of the material in the road construction. It will be found that most stones are suitable if their aggregate crushing value is not greater than 30 and, in the case of materials used at the road surface, provided the resistance to polishing is adequate. There are exceptions, especially in the limestone group, and the best guide is experience. The aggregate crushing value is determined by applying a load to a sample and measuring the percentage of lines formed. The extent to which an aggregate will become polished under the action of traffic is measured by means of an accelerated polishing procedure which simulates the effect of vehicle wheels on a road surface. Details of both these tests are presented in BS 812. It should be borne in mind that these considerations of crushing and polishing of aggregates are more significant in the design of major roads, where speeds, wheel loads and traffic volumes are high, than they are in the design of lightly trafficked roads in residential areas.

The flexible pavement

ASPHALTIC ROAD MATERIALS

The basic materials used in the construction of asphaltic pavements are course and fine aggregates, bitumen and tar as binding agents, and fillers of cement or limestone dust.

Coarse aggregate The surface of many asphaltic roads is smooth-textured and black in colour. Although this may provide the most satisfactory appearance in certain situations there are other settings where a different colour or texture would be more in keeping with the surroundings. The main way of obtaining a different surface is by the judicious use of course aggregate, either as a constituent of the wearing course mix or added to the surface by rolling in or as a surface dressing. This component then is of paramount importance when considering colour and texture.

Fine aggregate The function of the fine aggregate is, together with the binder and filler, to fill the voids formed by the particles of course aggregate. The proportion and grading of the fine material helps to determine the texture of the surface.

Course and fine aggregates are described above under the heading 'Materials of construction common to basic road types', on p. 222.

Filler This is the micro-aggregate which passes a BS No 200 mesh sieve. It is a fine inert mineral powder added to a bituminous binder in order to increase the bulk of the binder and adjust its properties. The addition of a filler increases the viscosity and thus reduces run-off of binder from aggregate.

Binder This material holds the particles of aggregate in place. When hot it acts as a lubricant, thus assisting compaction. Bituminous road binders are of two kinds, namely road tar and road bitumen. These materials have important qualities in common. They are both viscous and semi-solid, their degree of viscosity is dependent on temperature, and they both adhere to roadstone. Also both materials can be prepared in an extremely wide range of viscosities.

Tar (BS 76:1974 Specification for tars for road purposes) Tar is produced in Great Britain from the destructive distillation of coal. It is available in different grades, the two main categories being road tars for surface dressing and road tars for coated macadams.

Tars are cheaper than bitumens but they are less durable, especially in open textured wearing courses, since they weather more quickly; the reason being that it is easier for oxidation and polymerization to take place. Tars are more susceptible to changes in temperature so they harden more quickly than bitumens and become soft in hot weather. The resistance of tars to chemicals is greater than bitumens and thus they are not so affected by fuel spillage. However they are also resistant to solvents and so cleaning of equipment can be a problem.

Tar emulsions can be made but they are difficult to prepare and handle and so are very seldom used in Britain.

Bitumen (BS 3690:1970 Specification for bitumens for road purposes) This material is the highest boiling residue of crude petroleum oil and is usually manufactured in a vacuum distillation plant. Bitumen, obtainable in different grades, being more durable than tar is unlikely to need resurfacing before six or seven years has elapsed as opposed to four or five. This time may be longer depending on the type of surfacing and conditions of use. Cut-back bitumen is one which has been thinned down by the addition of a volatile solvent, kerosene or creosote, and is produced in order to provide a more workable material. Bitumen emulsions are also available. The manufacture of these involves another method of making the bitumen more fluid. They consist of bitumen dispersed in water with a stabilizing additive. Emulsions are sufficiently fluid to be used cold.

Although bitumen is obtainable from natural asphalt, that produced from petroleum by a refinery-distillation process constitutes by far the largest proportion used in Britain.

Pre-mixed bituminous materials

The two main types of premixed bituminous materials are coated macadams and asphalts. These materials consist of a mixture of mineral aggregates and a binder with, in some cases, the addition of a filler. They are manufactured in a plant and delivered to the site ready for use. The binder may be bitumen or tar and the term bituminous may refer to either. The aggregate strength in the asphalts is supplemented by means of a strong mortar whereas the coated macadams rely for their strength on the interlocking between the aggregate particles. Some asphalts are a type of concrete, the cementing agent being bitumen. For a definition of asphaltic concrete see the glossary of terms on p. 219. Asphalt contains a relatively hard bituminous constituent whereas in the coated macadam a cut-back or soft bituminous binder is used in some of the mixtures although with dense macadam medium viscosity grades of binder are employed. The difference between the various macadams is a function of the aggregate size and grading, the type and grade of binder employed, and the proportions of binder to aggregate. The same is true of the asphalts. The choice of type of macadam or asphalt used will depend primarily on the subsoil conditions, the traffic loading, the layer being designed, the method of construction and the availability and cost of plant and materials.

COATED MACADAMS Coated macadam is defined as a road material consisting of graded aggregate that has been coated with a tar or bitumen, or a mixture of the two and in which the intimate interlocking of the aggregate particles is a major factor in the strength of the compacted roadbase or surfacing. Thus tarmacadam is coated macadam in which the binder is wholly or substantially road tar and bitumen macadam is one in which the binder is wholly or substantially bitumen. The word 'substantially' is included to cover the cases in which a mixture of the two binders is used. There is always a preponderance of one.

Coated macadams are classified according to nominal size and texture. The nominal size is the size of the largest aggregate in the material. The surface texture is controlled by the fines content, in other words the amount of aggregate in the material which passes the 3.35 mm BS sieve. The surface texture is usually defined by one of the following three categories:

Open-textured—15% of fines or less
Medium-textured—about 25% fines
Close-textured or dense—about 35% fines.

The open and medium-textured types which are used for basecourse and wearing course construction, are made from less viscous binders than the dense types, and are used for less heavily trafficked roads. The dense types are available for wearing course, basecourse and roadbase construction and are suitable for all types of roads.

Two well established wearing course materials come under the heading of coated macadams. These are cold asphalt and dense tar surfacing.

Cold asphalt (BS 4987:1973) For the definition see the glossary of terms on p. 219. In spite of the word 'cold' in the title it is normal practice to use a bitumen of such viscosity that the material has to be laid and compacted while still warm.

Dense tar surfacing (DTS) (BS 5273:1975) For the definition see the glossary of terms on p. 219. Although DTS is included under the general heading of coated macadams, its strength depends partly on the mortar. Like asphalts it is classified according to the stone content of the mixture. This is either 35% with coated chippings rolled into the surface or 50%, which itself provides an adequate surface.

ASPHALTS These are materials consisting of a mixture of asphaltic cement and mineral matter. The most commonly used are rolled asphalt (hot process) and mastic asphalt.

Rolled asphalt (hot process), (BS 594:1973) For the definition see the glossary of terms on p. 219. The British Standard specifies hot process rolled asphalt with rock, gravel or slag aggregate. It deals with rolled asphalt laid as a wearing course, a basecourse or a roadbase for roads and other paved areas.

Rolled asphalts are categorized by their coarse aggregate content, in other words by the percentage of the total amount of aggregate retained on the No 7 sieve (2.36mm) and in practice the compositions can be reduced to a 'low' (about 30%) or increased to a 'high' (about 60%) coarse aggregate content.

The low coarse aggregate content mixtures are used solely for wearing courses and normally have coated chippings rolled into the surface. The high stone content mixtures may also be used for wearing courses but are now more usually limited to the basecourse and roadbase. In all cases the maximum size of the coarse aggregates is conditioned by the thickness of the layer being constructed.

The asphaltic cement in rolled asphalt may be petroleum bitumen, lake asphalt/bitumen mixture or a pitch/bitumen mixture.

Mastic asphalt (BS 1446:1973 and BS 1447:1973) For the definition see the glossary of terms on p. 219. The term 'asphaltic cement' is used in connection with mastic asphalts to include petroleum bitumen, lake asphalt and mixtures of the two. The main difference between the two specifications is that material to BS 1446 is made from natural rock asphalt, ie, a rock naturally impregnated with bitumen, whereas material to BS 1447 is cemented with a manufactured mixture of limestone powder and bitumen. Unlike the other types of pre-mixed bituminous materials, mastic asphalt for roadworks is not normally manufactured in one operation. The mastic (or mortar) part is manufactured in special mixers at a depot and cast into blocks. These are taken to the site, reheated and mixed with up to 45% of chippings after which the material is spread, usually by means of hand floats.

The material is normally used only as a wearing course, and on carriageways coated chippings are rolled into the surface to provide resistance against skidding.

The family tree showing types and uses of pre-mixed bituminous materials is shown in figure **9.7**, and Table 1 lists the macadam and asphalt materials specified in BS 5273, BS 4987, BS 594 and the Department of Transport's *Specification for Road and Bridge Works*. The position of the specified materials in the system of products can thus be easily seen.

TABLE 1 Macadam and asphalt materials specified in the Department of Transport's *Specification for Road and Bridge Works* **and in BS 5273, BS 4987 and BS 594**

Materials	Specification
Wearing course	
Coated macadams:	
14mm nominal size open-textured macadam	Open-textured bitumen macadam, clause 912; BS 4987
10mm nominal size open-textured macadam	Open-textured tarmacadam, clause 913; BS 4987
6mm nominal size medium-textured macadam	BS 4987
14mm nominal size dense macadam	Dense bitumen macadam, clause 908; BS 4987
10mm nominal size dense macadam	
Course cold asphalt★ (10mm nominal size medium-textured bitumen macadam)	Cold asphalt,★ clause 910; BS 4987
Fine cold asphalt★ (6mm nominal size fine-textured bitumen macadam and 6mm nominal size fine-textured tar macadam)	
Dense tar surfacing	Dense tar surfacing, clause 909; BS 5273
Asphalts:	
Rolled asphalt	Rolled asphalt, clause 907; BS 594
Basecourse	
Coated macadams:	
40mm nominal size open-textured macadam	Bitumen macadam, clause 905; BS 4987
20mm nominal size open-textured macadam	Tarmacadam, clause 906; BS 4987
40mm nominal size single course macadam	
40mm nominal size dense macadam	Dense bitumen macadam, clause 903; BS 4987
28mm nominal size dense macadam	
20mm nominal size dense macadam	Dense tar macadam, clause 904; BS 4987
Asphalts:	
Rolled asphalt	Rolled asphalt, clause 902; BS 594

Roadbase

Coated macadams:

40 mm nominal size dense macadam	Dense bitumen macadam, clause 811; BS 4987 Dense tarmacadam, clause 810; BS 4987

Asphalts:

Rolled asphalt	Rolled asphalt, clause 812; BS 594

Notes:

1 The clause numbers listed under Specification *refer to those in* Specification for Road and Bridge Works

2 Although in most cases the Department of Transport's Specification *refers to the British Standards, it carries provisos. In every case reference must be made to all the relevant documents*

★The cold asphalts are a medium-textured or fine-textured type of coated macadam

THE ASPHALTIC PAVEMENT

The pavement will normally consist of three main layers; the sub-base, the roadbase and the surfacing, as has been illustrated in **9.6**, above.

The sub-base

The functions of the sub-base can include those listed below although it may not be required to fulfil all these functions in a particular case.

1 The provision of a drainage layer to the road construction
2 Assistance in the load spreading function of the road
3 Initial protection to the sub-grade and provision of a preliminary running surface for the contractor's vehicles
4 Making up the total thickness required for the protection of the sub-grade from frost.

In some cases a sub-base is not required especially if the subgrade is sand or gravel.

The Transport and Road Research Laboratory's Road Note 29 includes a diagram which enables the thickness of material required for this layer to be determined, depending on the traffic loading and the California Bearing Ratio (CBR) of the sub-grade.

For roads under its jurisdiction the Department of Transport specifies the materials acceptable for the construction of sub-bases in its *Specification for Road and Bridge Works*; clause numbers in brackets refer to the clause numbers of that publication.

These materials include granular sub-base material Type 1, granular sub-base material Type 2, soil cement, cement-bound granular material, lean concrete, and wet lean concrete.

The materials are divided into three sub-base groups and these groups are related to traffic categories in table NG/1 of the Department of Transport's *Notes for guidance on the specification for road and bridge works*.[35]

a)   Granular sub-base material Type 1 (clause 803) must be crushed rock, crushed slag, crushed concrete or well burnt non-plastic shale. The grading limits are specified, as is the case for all the sub-base materials listed.

b)   Granular sub-base material Type 2 (clause 804) which may be natural sands and gravels as well as the materials listed for Type 1.

c)   Soil cement (clause 805). The materials permitted to be used for stabilization in the soil cement producing process are soil, chalk or pulverized fuel ash, a washed or processed granular material, crushed rock or slag, well burnt shale, spent oil shale or any combination of these. The required crushing strength is given. Soil cement is most practical when the existing ground is suitable for stabilization using in-situ mixing methods.

d)   Cement-bound granular material (clause 806). This

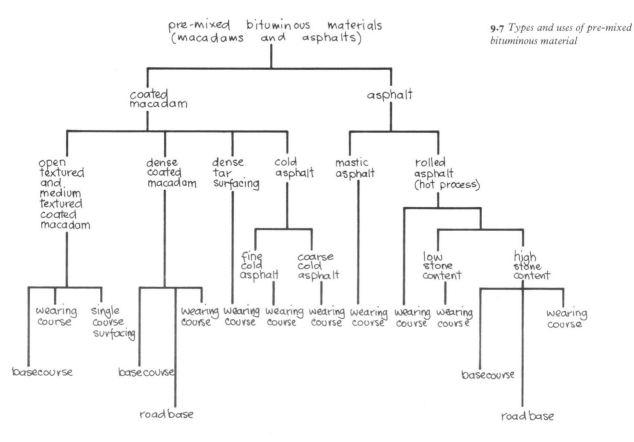

9.7 *Types and uses of pre-mixed bituminous material*

material is composed of a suitable granular material mixed with sufficient cement to achieve a prescribed crushing strength. For the granular constituent the Department of Transport's specification lists naturally occurring gravel-sand, a washed or processed granular material, crushed rock or slag or any combination of these.

e) Lean concrete (clause 807) consists of a graded aggregate mixed with a proportion of cement which is low by comparison with that used for most structural concretes. The aggregate is either coarse and fine aggregates batched separately or an all-in aggregate, having a maximum nominal size not exceeding 40mm nor less than 20mm. Crushing strength requirements are detailed in the specification.

f) Wet lean concrete (clause 815). The aggregate has the same specification as for cement-bound granular material. The mix must reach a specified crushing strength and the ratio by mass of cement to aggregate in the saturated surface dry condition must not be less than 1:20, unless approved by the engineer.

For the design of roads which need not comply with the Department of Transport's specification and which are not so heavily trafficked a number of other options exist. Provided they fulfil the general requirements of a sub-base as outlined above, the following unprocessed materials may be suitable.

Quarry waste or scalpings Natural grading with maximum size not greater than 100mm. It should not have more clay binding material than is required to hold the stones together.

Hardcore Clean broken stone, bricks, etc. Maximum size should not be more than 150mm.

Hoggin The mixture of sand, gravel and clay should not have more clay than is necessary to adequately bind the material. There must be no clay lumps.

Shale This material, which is taken from colliery tips, should be obtained from those tips which are known to be suitable and which do not contain any unstable elements.

The roadbase

The roadbase forms the main loadbearing thickness in the road construction. It may consist of unbound macadam or of aggregate bound with cement or with a bituminous binder.

The Department of Transport's specification permits the use of soil cement (clause 805), cement-bound granular material (clause 806), lean concrete (clause 807), wet mix macadam (clause 808), dry-bound macadam (clause 809), dense tarmacadam (clause 810), dense bitumen macadam (clause 811), rolled asphalt (clause 812), and any other approved roadbase material as described in the contract. Five roadbase groups are tabulated and table NG/1 relates the groups to traffic categories.

For lighter traffic loadings (less than 2.5 million standard axle loads), other materials which have been found suitable under the prevailing conditions may be considered. Whatever material is used it should not be susceptible to frost action.

A short description of each of the non-bituminous roadbases is presented below, a general description of the bituminous bound materials having already been given.

a) Soil cement (clause 805), cement-bound granular base (clause 806), and lean concrete (clause 807). It is possible to use these materials in flexible construction because they are weak in tension and crack under load. Soil cement and cement-bound granular bases are used for the more lightly trafficked roads whereas for those more heavily trafficked the lean concrete base may be adopted (see table 8/2 of specification). This entails a stronger mix and thus greater rigidity and better load distributing properties. With the cement-bound granular base a close network of cracks forms and the material acts like interlocking hardcore. In the case of the lean concrete larger cracks form at 3.6–4.5m spacing and an adequate covering of bituminous material is an essential part of the design. The composition of these materials for use in the roadbase is the same as that for use in the sub-base.

b) Wet-mix macadam (clause 808). This is a plant manufactured material. Crushed rock or slag is accurately graded, batched and mixed with a specified amount of water in order to produce a damp mix. The water ensures a minimum amount of segregation during transport and laying and also facilitates compaction.

c) Dry-bound macadam (clause 809). This material is not premixed, being formed in-situ as described below. Coarse and fine aggregate consisting of crushed rock or crushed slag are used. The coarse aggregate is laid by machine to a thickness within the range 75–100mm and rolled with two passes of a smooth-wheeled roller of specified weight. The fine aggregate is then mechanically spread to a thickness of 25mm and vibrated into the interstices of the coarse material. Spreading and vibrating of fine aggregate is repeated until no more will penetrate, at which stage it is rolled and the whole operation repeated until the total required thickness of roadbase has been constructed.

Surfacing

The surfacing may be either single or two course, the latter being composed of a basecourse and a wearing course.

The British Standard specifications for basecourse and wearing course asphaltic materials are contained in BS 5273:1975, Specification for dense tar surfacing for roads and other paved areas, BS 4987:1973 Specification for coated macadam for roads and other paved areas, and BS 594:1973 Specification for rolled asphalt (hot process) for roads and other paved areas. The Department of Transport specifies these materials in *Specification for Road and Bridge Works*, referred to above.

The recommended relationship between types of basecourse and wearing course and the traffic loading is shown in Table 2, which reproduces Table 4 of Road Note 29 and may be read in conjunction with Table 1 on p. 224, which lists the bituminous materials used in the road pavement, and with Table 4 on p. 230 in the section Summary of Asphaltic Pavement Surfaces, in order to obtain a general picture of the use of asphaltic road surfacings.

THE BASECOURSE The basecourse materials have already been briefly described above, under the heading Premixed Bituminous Materials on p. 223.

THE WEARING COURSE The wearing course materials, which are either coated macadams, dense tar surfacing, rolled asphalt (hot process), or mastic asphalt have also been briefly described above, under the heading Premixed Bituminous Materials, on p. 223.

BLINDING WITH BITUMINOUS GRIT The surface of newly compacted open-textured macadam may, if specified, be blinded by hand or mechanically with bituminous grit. The coated grit fills the voids in the surface and produces a more finely textured finish.

ROLLING IN COATED CHIPPINGS When the wearing course is fine cold asphalt (fine-textured bitumen macadam) the design may call for the rolling in of coated chippings. This will modify the surface texture. The size of the aggregate and the rate of distribution will have an influence on the effect produced. The following range of rates of distri-

bution are listed in BS 4987:

10mm chippings: 4–8 kg/m²
14mm chippings: 5–10 kg/m²
20mm chippings: 6–13 kg/m²

Coated chippings are also rolled into the surface of dense tar surfacing, mastic asphalt and rolled asphalt. With the latter, 14mm and 20mm nominal size are specified in BS 594.

SURFACE DRESSING Surface dressing is the treatment whereby a film of binder is sprayed over the road surface and is followed by the application of a single layer of stone

TABLE 2 Recommended bituminous surfacings for newly constructed flexible pavements (see Note 1)

Traffic (cumulative number of standard axles)			
Over 11 milions (1)	**2.5–11 millions** (2)	**0.5–2.5 millions** (3)	**Less than 0.5 million** (4)
Wearing course (crushed rock or slag coarse aggregate only) **Minimum thickness 40 mm** Rolled asphalt to BS 594 (pitch-bitumen binder may be used) (Clause 907)		**Wearing course** **Minimum thickness 20 mm** Rolled asphalt to BS 594 (pitch-bitumen binder may be used) (Clause 907) Dense tar surfacing to BTIA Specification (Clause 909) Cold asphalt to BS 1690 (Clause 910) (See Note 4) Medium-textured tarmacadam to BS 802 (Clause 913) (to be surface-dressed immediately or as soon as possible—see Note 4) Dense bitumen macadam to BS 1621 (Clause 908) (see Note 4) Open-textured bitumen macadam to BS 1621 (Clause 912) (see Note 4)	**Two-course** (a) **Wearing course—** **Minimum thickness 20 mm** Cold asphalt to BS 1690 (Clause 910) (see Note 4) Coated macadam to BS 802, BS 1621, BS 1241 or BS 2040 (Clause 913, 912 or 908) (see Notes 2 and 4) (b) **Basecourse** Coated macadam to BS 802, BS 1621, BS 1241 or BS 2040 (Clause 906 or 905) (see Note 2) **Single course** Rolled asphalt to BS 594 (pitch-bitumen binder may be used) Dense tar surfacing to BTIA Specification (Clause 909) Medium-textured tarmacadam to BS 802 (Clause 913) (to be surface-dressed immediately or as soon as possible—see Note 4)
Basecourse **Minimum thickness 60 mm** Rolled asphalt to BS 594 (Clause 902) (see Note 2) Dense bitumen macadam or dense tarmacadam (crushed rock or slag only) (Clause 903 or 904)	**Basecourse** Rolled asphalt to BS 594 (Clause 902) (see Note 2) Dense bitumen macadam or dense tarmacadam (Clause 903 or 904) (see Note 3)	**Basecourse** Rolled asphalt to BS 594 (Clause 902) (see Note 2) Dense bitumen macadam or dense tarmacadam (Clause 903 or 904) Single-course tarmacadam to BS 802 (Clause 906) or BS 1241 (see Notes 2 and 5) Single-course bitumen macadam to BS 1621 (Clause 905) or BS 2040 (see Notes 2 and 5)	Dense bitumen macadam to BS 1621 (Clause 908) (see Note 4) 60mm of single-course tarmacadam to BS 802 (Clause 906) or BS 1241 (to be surface-dressed immediately or as soon as possible—see Note 4) 60mm of single-course bitumen macadam to BS 1621 (Clause 905) or BS 2040 (see Note 4)

Notes:

1 The thicknesses of all layers of bituminous surfacings should be consistent with the appropriate British Standard Specification

2 When gravel, other than limestone, is used, 2% of Portland cement should be added to the mix and the percentage of fine aggregate reduced accordingly

3 Gravel tarmacadam is not recommended as a basecourse for roads designed to carry more than 2.5 million standard axles

4 When the wearing course is neither rolled asphalt nor dense tar surfacing and where it is not intended to apply a surface-dressing immediately to the wearing course, it is essential to seal the construction against the ingress of water by applying a surface dressing either to the roadbase or to the basecourse

5 Under a wearing course of rolled asphalt or dense tar surfacing the basecourse should consist of rolled asphalt to BS 594 (Clause 902) or of dense coated macadam (Clause 903 or 904)

Source: *Table 4, Transport and Road Research Laboratory, Road Note 29, A guide to the structural design of pavements for new roads, HMSO, London, 1970*

chippings which are then rolled. This is a low cost and simple form of maintenance but it also provides an excellent method of enhancing the texture and colour of an asphaltic road surface. The designer can choose from a wide range of stone types especially in the case of roads carrying low volumes of traffic at relatively low speeds. He can also choose from a range of stone sizes, so that both the texture and colour of the surface can be designed to best suit the surroundings.

Surface dressings should conform with Road Note 39, Recommendations for road surface dressing, produced by the Transport and Road Research Laboratory.[36] Dressings which are applied in accordance with this document should last at least five years before further treatment is required and in some cases they may last much longer.

The main engineering purposes of surface dressing are:

1 To seal the surface of the road and thus protect it from deterioration due to the ingress of water

2 To arrest or prevent the start of disintegration from the influence of traffic

3 To improve the skid-resisting qualities of the surface.

All types of road can be successfully treated in this way whether they are lightly or heavily loaded and the treatment is applied to new roads as well as existing ones.

The specification for the surface dressing will depend on the type of road surface to be dressed and on the volume of traffic to be sustained. Embedment of the chippings is a function of the surface hardness and the weight of traffic. Road Note 39 describes a range of surface hardnesses from very hard to very soft. For the design of surface dressing, traffic is considered in terms of the number of commercial vehicles per day in the lane under consideration, the commercial vehicle being defined as a vehicle of laden weight greater than 1.5 tonnes (Mg).

Table 3, which lists the recommended nominal size of chippings for combinations of surface type and traffic loading, is reproduced from Road Note 39.[37]

TABLE 3 Recommended nominal size of chippings (mm)

| Type of surface | Lane traffic category with approximate number of commercial vehicles currently carried per day | | | | |
	(1) Over 2000	(2) 1000–2000	(3) 200–1000	(4) 20–200	(5) Less than 20
Very hard	(10)	10	6	6	6
Hard	14	14	10	6	6
Normal	20*	14	14	10	6
Soft	X	20*	14	14	10
Very soft	X	X	20*	14	10

Note:

The size of chipping specified is related to the mid-point of lane traffic category ranges 2–5; lighter traffic conditions may make the next smaller size of stone more appropriate

* *At the discretion of the engineer, 20 mm chippings may be used for remedial treatment where traffic speeds are low. Very particular care should be taken when using 20 mm chippings to ensure that no loose chippings remain on the surface when the road is opened to unrestricted traffic as there is high risk of windscreen breakage*

X *Unsuitable for surface dressing*

Source: *Transport and Road Research Laboratory, Road Note 39, Recommendations for Road Surface Dressing, HMSO, London, 1970*

The choice of chipping size to suit the amount of traffic and hardness of road surface is illustrated in figure **9.8**, reproduced from figure 2 of Road Note 39.

The best results are obtained with single-sized chippings and so the stone or gravel should comply with BS 63 Single-sized roadstones and chippings, or BS 1984 Single-sized gravel aggregates for roads.

Rounded stones or gravel are readily available over large areas of Great Britain and therefore are very useful aggregates for surface dressing. To ensure stability particular care must be taken to use the correct quantity of binder.

If conditions over the early part of the life of the dressing are good any chippings having the correct size and satisfactory mechanical properties can be used to make a good dressing.

It is strongly recommended that compaction is carried out using slow-moving, rubber-tyred rollers. Steel rollers tend to crush the chippings. If it is necessary to use steel rollers the lightest ones available should be employed and in no case should the weight exceed 8 tonnes (Mg).

The binders used include road tar, cut-back bitumen, bituminous emulsions, and tar-bitumen blends. The quantity of binder in relation to the quantity of aggregate is critical. The binder should rise up the side of the aggregate to a sufficient extent to ensure good adhesion, but not so far as to produce a smooth surface. This principle is illustrated in **9.9**, which is reproduced from figure 12.1 of *Bituminous Materials in Road Construction*.[38] Although prepared to illustrate the spread of binder on old road surfaces, the same considerations apply to new surfaces. A good surface dressing gives a shoulder-to-shoulder coverage after compaction by traffic. The binder should be completely covered by the chippings.

The surface dressing of new open-textured coated macadam should be carried out after pressure from the traffic has sufficiently reduced the surface voids. If the surface dressing is to be applied within a year of laying the open textured bituminous macadam, then the voids should be filled with coated grit or fine cold asphalt.

When surfaces are permeable (as a coarsely graded coated macadam may be) the recommendations for size of chippings in Table 3 above do not apply. For porous surfaces a correction must be applied and the size of chipping given in Table 3 should be reduced to the next size smaller, though never smaller than 6 mm.

SUMMARY OF ASPHALTIC PAVEMENT SURFACES A summary of the textures and surface finishes associated with coated macadam and asphalt is presented in Table 4 together with other relevant information.

Unbound surfaces

Unbound flexible surfacings are laid without binders, invariably on an unbound roadbase. The pavement, then, consists of a roadbase and sub-base when necessary, with an unbound surface material superimposed. The roadbase usually consists of clinker or hardcore.

Gravel—general

Gravel is obtained from pits or river beds or from crushed stone. A great variety of types are available, producing different coloured surfaces and different textures. Gravel is suitable for lightly trafficked areas where violent braking/accelerating are unlikely and where consistent maintenance can be expected such as driveways to dwellings.

9.8 *Use of different size chippings to offset embedment produced by traffic forces in substrates of different hardness*

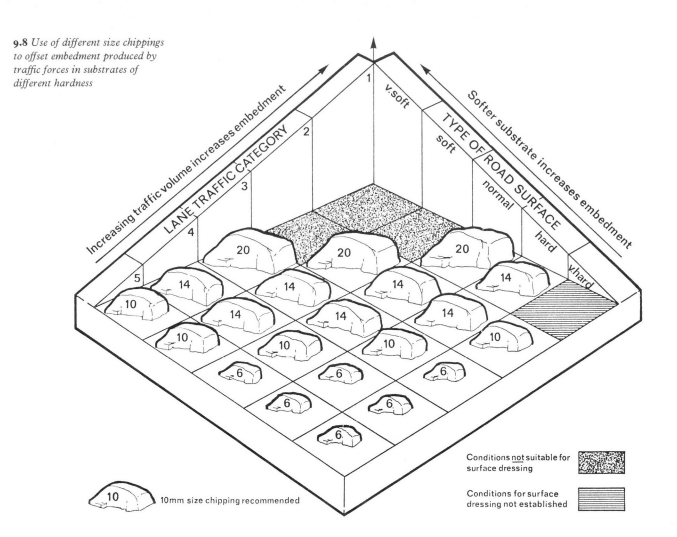

Conditions <u>not</u> suitable for surface dressing

Conditions for surface dressing not established

10mm size chipping recommended

9.9 *Diagram illustrating the rate of spread of binder related to the chippings and the old road surface*

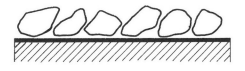

a *Normal proportion of binder on a smooth and impervious surface; rough or porous surfaces will absorb binder and more will be required to hold the chippings*

b *Chippings displace binder; if the surface is very hard and the chippings are not bedded into it then more binder is required to hold them*

c *Compaction and embedment still further displace binder up and around chippings; any further pushing of the chippings into the old surface will result in 'fatting-up'**

9.9

d *Rounded chippings leave greater voids and do not interlock, hence more binder is required to hold them*

*The term 'fatting-up' refers to the condition where too much binder has been used, with the result that it tends to rise to the level of the top of the aggregate and create a smooth surface.

TABLE 4 Textures, surface finishes and other characteristics of asphaltic pavements

Classi-fication	Material	Texture	Surface finish	Use	Aggregate	Method of laying	Interval between mixing and laying	Laying temper-ature	Comments
Coated macadams: warm and cold laid materials	Tar macadam	Open to medium Medium to fine Open to fine	As laid or coated grit blinding or surface dressing	Open and medium texture generally restricted to 'lesser roads'; dense macadams can be used under heavier traffic conditions	Rock, slag, gravel	Normally machine Can be hand spread	Normally up to 24 hours depending on proportion of flux in mix; macadams with little or no flux need to be laid as soon as possible, ie, while hot	Cold, warm, hot	Tar content makes it more resistant to oil droppings so frequently used for car standings
	Bitumen macadam	Open to medium Medium to fine Open to fine	As laid or coated grit blinding or surface dressing		Rock, slag, gravel	Normally machine Can be hand spread	Can be increased to some days by special flux	Cold, warm, hot	Can be used for car standings if oil dropping is not severe
	Dense tar surfacing	Fine Open to fine Open to fine Open to fine	As laid or surface dressing or coated chippings, coated gravel rolled in or uncoated chippings, uncoated gravel rolled in	All classes of road but normally on less heavily trafficked roads	Rock, slag	Normally machine Can be hand spread	Summer conditions 3 hours; winter conditions 2 hours	Hot	Increasing texture from tar weathering Resistant to softening by oil droppings Uncoated chippings only used on lightly trafficked roads, eg, housing access roads
	Cold asphalt	Medium to fine	As laid or coated chippings, coated gravel rolled in	Not used on main roads	Rock, slag	Normally machine Can be hand spread	Some months if suitably fluxed	Cold or warm	
Asphalts	Rolled asphalt (hot process)	Fine Open to fine Open to fine Open to fine	As laid or surface dressing or coated chippings, coated gravel or uncoated chippings, uncoated gravel rolled in	Heavily trafficked roads	Rock, slag, gravel	Normally machine Can be hand spread	1½ to 2 hours	Hot	Cheapest and most common of hot asphalts If stone content less than 45%, chippings usually rolled in to surface; if stone content more than 45% not necessary to roll in chippings Uncoated chippings or gravel only used on lightly trafficked roads, e.g. housing access roads
	Mastic asphalt	Fine, smooth Open to fine Open to fine Sand paper texture	As laid or surface dressing or coated chippings, coated gravel or sanding or indenting (by crimping roller)	Not normally used on main carriageways; usually restricted to special uses, e.g. bus bays, high indentation risk areas	Rock, lime-stone	Normally machine Can be hand spread	Reheat on site	Hot	Coated chippings usually used for roads Crimping roller usually only used for paths

Notes:
1 With surface dressing a variety of texture and colour is available depending on the size, shape and colour of the aggregate
2 With coated chippings a variety of texture is available depending on the size and shape of aggregate. Since coating will wear off due to traffic, colour can also be determined
3 With rolled in, uncoated stone the appearance is affected by the size, shape and colour of the aggregate and also by the rate of spread
4 The longest life as a wearing surface without subsequent treatment is generally considered to be obtained with mastic asphalt, compressed natural rock asphalt and hot rolled asphalt. The most economic selection is dependent upon local circumstances. Other surfaces do not usually give so long a life as a wearing course, but their service can be extended by periodic surface treatment

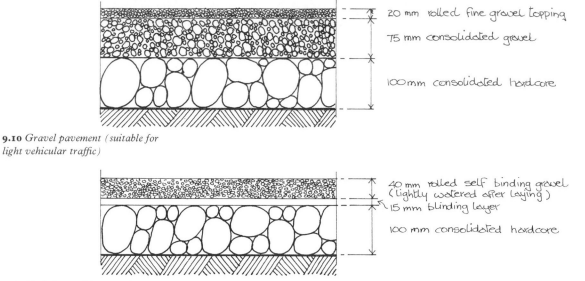

20 mm rolled fine gravel topping

75 mm consolidated gravel

100mm consolidated hardcore

9.10 *Gravel pavement (suitable for light vehicular traffic)*

40 mm rolled self binding gravel (lightly watered after laying)

15 mm blinding layer

100 mm consolidated hardcore

9.11 *Sealed gravel pavement (suitable for light vehicular traffic)*

Unsealed gravel

This is an excellent material for its informal and precinct-like character. It is especially useful for running surfaces adjacent to trees since it enables water and air to reach the roots. The surface should be treated with weed killer every spring and given adequate falls to avoid 'puddling' and consequent formation of moss.

The gravel is laid loose on a roadbase layer of consolidated hardcore or other suitable material and then rolled. A 10mm blinding layer may be applied on top of the roadbase although this might not be necessary. A wearing course of finer gravel may be added or a covering of fine grit rolled into the surface. Heavy rolling is normally required at all stages of construction (8–10 ton roller). A typical cross section is shown in **9.10**.

Sealed gravel

Some gravels contain a mixture of clay, sand and stone which imparts self-binding and sealing properties. When watered and rolled during laying the constituent elements bind together and form a firmly cemented surface. This material is suitable for lightly trafficked areas. It must be laid to falls and shrinkage cracks should be guarded against. Colours vary according to locality. Breedon gravel is obtainable in yellow/browns and reds; self-binding granite dusts come in pink/greys. White dust from lead (grey) and copper (coppery) are sterile and inhibit weed growth.

The gravel is laid on a blinded layer of hardcore and rolled dry to the required thickness. It is then sprayed with water and rolled until a slurry forms on the surface. This will be washed away by the first rain. As with unsealed gravel, an 8–10 ton roller should usually be employed. A typical cross section is indicated in **9.11**.

The specification varies depending on loading and local conditions. A typical specification for Breedon gravel is:

Breedon gravel finish for entrance roadways and drives (construction for foot traffic is shown by the figures in brackets).

The Breedon gravel surfacing for entrance roadways and drives, external paths and pavings, as indicated on the drawings, is to be composed and laid as follows:

Provide, spread and level over previously prepared

foundations good screened pit or river gravel to pass a 50mm (37mm) ring with sufficient hoggin for binding, and water and well roll with an 8–10 ton roller to specified levels and cross-section to a consolidated thickness of 62mm (50mm). Finish with Breedon red/buff/golden amber gravel obtained from the Breedon and Cloud Hill Lime Works Ltd, Breedon-on-the-Hill, Derby. Spread evenly to a thickness of 50mm (37mm) and dry roll until consolidated; afterwards roll wet by sprinkling water from a rose onto the roller and continue rolling until the surface of the gravel is covered with a wet slurry.

The rigid pavement

The rigid pavement normally consists of an in-situ slab and a sub-base. The principle structural element is the slab. It fulfils the function of the surfacing and the roadbase of the flexible pavement.

THE SUB-BASE

The functions of the sub-base include:

1 Forming a working surface on clays and silts, facilitating accurate control of slab thickness
2 Enabling work to proceed during wet weather without damage to the sub-grade
3 Preventing mud-pumping on clays and silts
4 Insulating frost susceptible soils.

Base materials should be hard, durable, chemically inert, suitably graded and capable of compacting to a high density. When compacted they should not be susceptible to appreciable frost heave or to shrinkage, swelling or loss of stability resulting from changes in moisture. Suitable materials may include the following (clause numbers refer to the Department of Transport's specification):

Granular materials (clauses 803, 804) including suitably graded or self-binding gravel, crushed stone or crushed slag. Selected well-burnt colliery shale may be used if local experience is satisfactory.

Soil cement (clause 805)

Cement-bound granular material (clause 806)

Lean concrete (clause 807)

Wet lean concrete (clause 815)

The ingredients of the slab are coarse aggregate, fine aggregate and cement. It may or may not require reinforcement depending on the design approach. Information on aggregate is given on p. 222 under the heading 'Materials of construction common to basic road types'. The British Standard Institution specification for both coarse and fine aggregate from naturally occurring materials is detailed in BS 882, 1201: Part 2: 1973, Specification for aggregates from natural sources for concrete (including granolithic). The cement is normally ordinary Portland cement complying with BS 12: Part 2: 1971, Portland cement (ordinary and rapid hardening).

The thickness of the slab and sub-base are functions of the quality of the sub-grade and the traffic loading. Thicknesses taking account of these factors are given in Road Note 29. The mix should be sufficiently workable to allow for thorough compaction; have adequate compressive and flexural strength; be dense to resist the abrasive and impact action of traffic; finished with an even surface to give a good riding quality, and with a texture which minimizes the possibility of skidding. Compaction without segregation is of utmost importance for strength, as 5% of voids can reduce strength by about 30%, and 10% voids by 60%. Vibrating machines compacting from the surface are normally adequate for slabs up to 300mm thick. The principal cause of damage can be the use of salt for snow or ice removal in winter. This is virtually eliminated by the use of:

1 An air entraining agent in the top 60mm of concrete to give 3–6% air in the form of small bubbles aided by

2 Free water/cement ratio less than 0.55 by weight.

When concrete is laid in winter it should be protected from frost for the first few days. If it can be maintained at a temperature of at least 4.4°C for three or four days it can usually be considered safe. It is most important to ensure that the subgrade and sub-base do not freeze.

Air entrained concrete should be used either for the full depth of the slab or for at least the top 50mm. (When the slab is reinforced it may be convenient to use air entrained concrete for the full depth above the reinforcement.)

JOINTS

The appearance of concrete roads can be marred by badly detailed joints. Black sealing compounds are sometimes extruded and form ragged black lines across the surface. Overfilled joint grooves sealed flush with the road surface are a prime cause of this. These lines are emphasized by the, usually, light colour of the concrete. The spalled edges of joints or cracks in the slab due to inadequate joint design also cause unsightly lines.

Proper design, good detailing and careful construction of these elements are of primary importance in achieving a satisfactory appearance with a concrete road slab. Careful consideration should be given to the colour of the material under consideration as a sealant and also to its long term performance. It should not be chosen on the basis of cost alone or even on the basis of long term performance and cost only.

Specifications for joints in concrete pavements are given in the Department of Transport's *Specification for Road and Bridge Works*. Their design is treated in Road Note 29 and also in the Transport and Road Research Laboratory's *The Design and Construction of Joints in Concrete Pavements*.[39] Joints in a concrete road slab may be transverse or longitudinal. There are four types of transverse joint: expansion joints, contraction joints, warping joints and construction joints.

The purposes of joints are to allow for expansion, contraction and warping of the concrete slab due to changes in temperature and for contraction of the slab due to loss of moisture during the curing period. As well as allowing for these movements the joints must:

a) Maintain the riding quality of the road

b) Prevent water from penetrating below the slab and prevent the ingress of grit and stone

c) Allow an economic construction programme

d) Not produce a significant weakening of the slab.

Expansion joints (transverse) provide a gap in the slab, filled with compressible material, thus allowing for thermal expansion and contraction. These joints also allow for warping of the slab.

Contraction joints (transverse) are essentially a break in the concrete which permits contraction due to a fall in temperature and due to shrinkage of the slab in setting. Contraction joints are constructed at closer spacings than expansion joints. Some warping can occur at these joints.

Warping joints (transverse) are also breaks in the slab but ones in which differential vertical movement across the break, or opening of the joint, is prevented by the addition of tie-bars. These joints also allow some angular movement but they do not allow for contraction of the concrete.

Construction joints occur at positions in the slab where construction has had to be interrupted at a position other than that where joints had been planned as part of the design. These are to be avoided. Each pour should be arranged to finish at one of the planned joints.

Longitudinal joints are used when the slab width is in excess of 4.5m. They are designed to cope with transverse warping and with uneven settlement of the sub-grade. These joints also enable the road to be constructed in convenient widths. They are usually constructed with a similar form to warping joints.

Expansion joints (transverse)

The ends of the slab should be square and vertical and the gap should not at any point be less than the specified width. The compressible filler in the gap usually starts at the bottom of the slab and extends to 25mm from the top. Materials used include fibreboard, knot-free softwood and cork-based board. The top 25mm is filled with a sealing compound to prevent the ingress of water, grit or stones. The dowel bars across the joint are bonded to the concrete on one side but on the other the bond is broken by painting the bars with a suitable compound (the Department of Transport's Specification defines this material) and expansion is allowed by placing a cap containing compressible material on the end of each bar, **9.12**.

Contraction joints (transverse)

These are formed by positioning a groove in the surface of the concrete thus determining the position of the cracks and ensuring that they are not visible from the surface, **9.13** and **9.14**. The groove can be formed during construction or

cut into the concrete surface, by saw, after setting.
Dowel bars are required if the slab thickness exceeds 150mm.

Longitudinal joints

Although not as important as in transverse joints, it is desirable that loads are transferred from one side to the other of this type of joint. It is therefore desirable that tie bars are incorporated. These will also prevent the opening of the joints.

There are a variety of formers used for this type of joint, three being shown in **9.15**, **9.16** and **9.17**.

Construction joints

If the use of this type is unavoidable it should be constructed with a sealing groove and the reinforcement should be carried across the joint. If the sub-grade is poor,

tie bars may be required, **9.18**.

Sawn joints

Grooves in the surface of the concrete are often made today with water cooled saws. Although these may be more costly than preformed joints they are labour saving and provide a better finish.

Sealing grooves

The prepared grooves at the joints are filled with a poured sealing compound or a compression seal. The former type may be hot-poured or cold-poured. Materials used include rubber and bitumen mixtures, oxidized bitumen compounds and straight run bitumen compounds containing fillers and resinous material. Preformed compression seals may be made from bituminous material or a number of synthetic ones including synthetic rubbers.

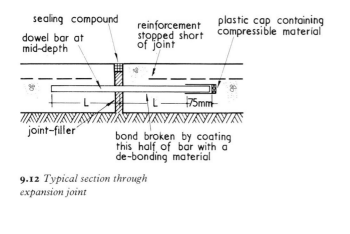

9.12 *Typical section through expansion joint*

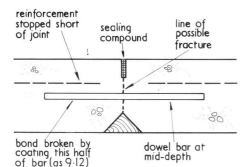

9.13 *Transverse contraction joint—fracturing controlled by slot and batten*

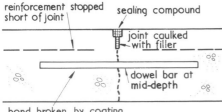
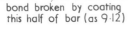

9.14 *Transverse contraction joint—fracturing controlled by deep slot*

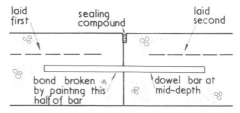

9.15 *Longitudinal butted construction joint*

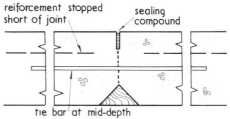

9.16 *Longitudinal joint—fracturing controlled by slot and batten*

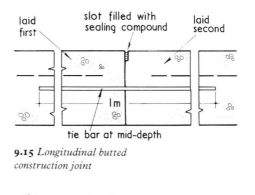

9.17 *Tongued and grooved longitudinal construction joint (unsuitable for slabs less than 200mm thick)*

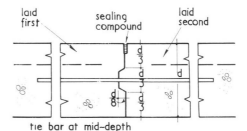

9.18 *Transverse butted construction joint*

Initial finish

In all in-situ concrete road slabs the initial finish will smooth out any irregularities and establish the surface at the final levels. This is achieved by careful pouring, screeding, and compacting with a compacting beam, and then passing over the surface with a skip float. Full compaction is essential for good performance and is usually carried out with a single or double beam compactor fitted with a vibrator unit, **9.19**. The vibrating process brings moisture to the surface and this should be removed in the course of the finishing process to ensure a hard and durable surface to the cured concrete.

After the initial finish the surface should be correctly level within the limit of 2 to 3mm in 600mm.

Final in-situ finishes—aggregate not exposed
Floated and trowelled finishes   Hand wood floating will give an attractive textured surface and is suitable for use over small areas where traffic is light. Hand floating gives a light ripple texture which is best applied in a series of overlapping areas.

Successive hand trowelling with steel trowels is a traditional method of obtaining a smooth hard surface. The operation should be carried out by a skilled worker to achieve good results. Surplus mortar should be removed during the process and at least a one hour gap should be allowed between each trowelling to allow surface water to evaporate. Power trowelling will also produce a hard smooth finish. In this process surplus mortar is not removed but is trowelled into the surface. This means that the timing of the operation is critical since surface moisture should have evaporated and the concrete must be set sufficiently to support the machine and operator and yet not be so hard that the mortar cannot be worked into the surface. A hard surface is produced by a series of trowellings with the blades at increasingly steep angles.

Bristle broom finish   Texture can be produced by brushing the surface of the concrete before it has hardened. After compaction a suitable preparation is the application of a scraping straight edge. This is a simple hand tool which is drawn across the surface rendering it level and smooth. This operation should be carried out a few minutes after compaction. The texturing is applied by drawing a broom across the surface when the concrete is still plastic but after the initial wet sheen has gone.

The appearance of the texture is a function of the type of bristles and the depth of the indentations. When a soft or medium bristled broom is used the depth of the texture will not exceed 0.5mm. This may be suitable for footpaths and lightly trafficked surfaces for vehicles but with heavy traffic the texture will soon become worn. Soft brooms with 450mm wide heads and 150mm long bristles can be used to give two degrees of texturing by applying normal or heavy pressure.

Bass brooms or wire brooms, **9.20**, are more appropriate for surfaces to be used by vehicles. Wire brooms are available with 450mm wide heads, spring steel tape bristles 0.3mm thick, 1.25mm wide and about 100mm long, arranged in two or four rows of tufts. The most satisfactory broom type is that with two rows of spring steel tapes. This is specified in the Department of Transport's *Specification for Road and Bridge Works*. Indentations running across the road width will be best for drainage but other arrangements may be desired. A panelled affect, for example, is easily produced by successive areas brushed at right angles to each other. Patterns can also be produced by smoothing out edges and stripes with a float after brushing has been carried out. The effect of particular orientations of texture in skid resistance should be considered where appropriate.

Tined finish   To produce deeper parallel grooves than a brushes finish a tined rake is sometimes used, **9.21**. However this should only be used on very lightly trafficked surfaces since the texture is not well compacted. Rakes can be designed to indent a variety of shapes in the surface. They should be constructed in such a way that they force aggregate into the ridges which should not just be formed of laitence. A marketed garden rake with curved rubber tines can be used to produce this type of heavier textured finish.

The spacing of the tines is important. 12mm is the lower limit, 15mm gives good results and 18mm is probably too large a spacing. The same possibilities regarding the creation of pattern exist with tined as with brushed finishes but the former will be more durable. Because of the deep grooves particular attention should be given to drainage.

Tamped finishes   Hand compaction has been carried out in the past by using the edge of a long board as a punner on the surface of the concrete. At the present time a much heavier board with a vibrator attached is normally employed for this work. In this case a second, lighter board must be used to produce the finish by tamping the surface with the edge of the board to cause closely spaced, parallel, small undulations or ribs on the surface, **9.22**. These are commonly placed at right angles to the road centre line and result in an attractive, well draining and skid resistant surface. The finish should be applied within half an hour of compaction.

Alternative finishes can be produced using a board with small projections located along its edge. This board is dragged along the surface thereby forming a grooved finish, the pattern being determined by the shape and spacing of the projections, **9.23** and **9.24**. The position and level of the board is guided by templates fixed to the side forms. A concrete mix with a high proportion of sand should be used and with some types of profiled screed board it is necessary to place a layer of mortar on top of the fresh concrete. It is then the mortar which is indented. If this is not done these boards will tear the surface. It is important to use the correct thickness of mortar when obtaining the finish in this way. However such a procedure will result in a less strong surface and should only be used in areas of light traffic.

Rolled finishes   These produce a rippled effect, **9.25**, which can help to disguise changes in texture or colour especially if a strong pattern is used. The ripples are formed by the mortar being drawn into ridges. The roller diameter should be large, say 150mm, and wider rollers are likely to produce better results than narrow ones. With the latter a ridge of mortar tends to form at the roller edge. The roller should also be sufficiently stiff not to deflect.

For texturing the surface of plastic concrete two types of

9.19

9.22

9.20

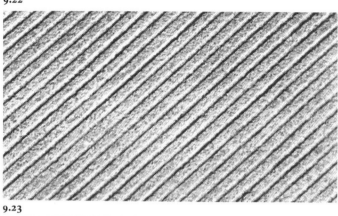

9.23

9.21

9.24

9.25

9.27

9.26

roller are available. One comprises a pair of rollers 127 mm in diameter and 1.1 m wide with expanded metal on the surface, **9.26**. The other is heavier and is constructed with 1 drum of 300 mm diameter and 3.5 m long to which a choice of patterned rubber or expanded metal sheets can be fixed depending on the finish required. The second roller is considered to give the best results and the most effective rubber wrapping for the drum gives a 5–6 mm deep ribbed pattern. Two types of expanded metal sheet can be used. The general purpose one is a diamond mesh with 30 × 12 mm openings and the other is similar but flattened; the latter gives shallower and less uniform results.

Machined finishes The concrete slab can be textured after setting by the use of mechanical groovers. The methods at present available range from the use of reflex hammer motors to tungsten carbide-tipped flails or tines. The cutting speed is largely dependent on the type of aggregate used and varies from 8–10 m a minute for limestone aggregates to 4–5 m a minute for flint aggregate.

Final in-situ finishes—aggregate exposed
Brushed and sprayed After the concrete has been poured, compacted and the surface levelled the aggregate can be

exposed by brushing and spraying with water. This process exposes the true nature of the concrete and, by careful selection of the colour, size and shape of the aggregate, finishes which are attractive in texture and colour are achieved.

After the concrete is levelled it should be brushed with a soft broom to remove any surface laitence then covered with waterproof sheeting to prevent too rapid drying out and left until all surface mortar can be removed by brushing and spraying without disturbing the surface aggregate. The correct time for the brushing and spraying is normally between two and six hours after pouring. The material round the aggregate should only be removed to a depth sufficient to reveal the colour and texture of the stone and should never exceed half the depth of the aggregate. Every care should be taken to maintain a uniform degree of exposure and to avoid disturbing the aggregate. Stiff bristle or wire brushes should be used with plenty of water and all mortar should be removed from the exposed surfaces of the pieces of stone. Finally the surface should be sprayed and lightly brushed to remove all traces of mortar.

If, after the final spraying and brushing, any cement bloom remains on the exposed aggregate this can be removed by

washing with a 10% solution of commercial hydrochloric acid. This treatment should not be carried out until at least three and preferably seven days after the aggregate is exposed. Afterwards the surface should be thoroughly washed down with water to remove all trace of the acid. The time period during which washing and brushing can be carried out after pouring may be extended to up to 24 hours by the use of a retarder. This should be applied by spraying, making sure that an even distribution occurs otherwise the final results will not be satisfactory. Care must be taken to ensure that all traces of the retarder are washed away during the process of exposing the aggregate. Producing concrete with a uniform exposed aggregate finish requires skill and attention to detail at all stages of design and construction. The mix must be correctly designed, the batching of concrete must be accurately controlled, placing and compaction must be carried out in a consistent manner and the process of exposing the aggregate should be undertaken by skilled men who are able to produce uniform results. The exposure of the aggregate will make apparent any flaws in the concrete due, for example, to inconsistent batching or excessive delays during the process of laying the concrete. Nevertheless the attractive finish and the superior weathering characteristics amply reward the extra effort required.

Integral aggregate finish Course aggregate producing the required colour and texture will sometimes be more expensive than that giving the required concrete strength. In this case it may be desirable to use a two layer construction, the bottom layer consisting of a normal mix and the top layer being 50mm thick, air entrained and incorporating the special aggregate to be exposed. Aggregate in the range 4–8mm should be omitted from the top layer mix. In this form of construction the second layer should be poured within half an hour of the first layer in order to ensure a monolithic action.

Broadcast (scatter) finish Although the normal method of obtaining an exposed aggregate finish with in-situ slabs is by the brushed and sprayed technique another method is sometimes used. This entails scattering a natural or mineral aggregate, usually mixed with cement, over the surface of the poured concrete, **9.27**. The pre-mixed material is scattered to attain a specified distribution, immediately after the compaction of the slab. It is then compacted with a power float and finished by trowelling. The process, to be successful, must be carried out by skilled workers and a firm with the necessary expertise should be engaged.

In cases where an exceptionally high abrasive strength is needed a special topping mix is sometimes incorporated. Similarly a sand/cement topping may be used for decorative reasons depending on the surface texture required.

There are three methods of constructing the toppings: monolithic, bonded and unbonded construction. (They should always be used with the appropriate structural grade of concrete.) The recommended thickness and bay areas for these three methods is given in Table 5 below.

Monolithic construction is the cheapest method as well as the one likely to give the best results. The topping is 12–20mm thick and the structural slab is finished at a level sufficiently low to accommodate the topping thickness.

This is spread not more than three hours after the main part of the slab has been constructed so that it is completely monolithic. It is then compacted and finished. The topping becomes part of the structural slab.

TABLE 5 Recommended thickness and bay areas for high-strength concrete toppings

Type of construction	Thickness of topping (mm)	Bay areas	Comments
Monolithic	12–20 (allow 3–4mm surcharge)	As for structural slab	Topping laid within 3 h of finishing slab
Bonded	20–40 (allow 4–8mm surcharge)	Maximum width 4.5m Maximum area 25 m² Length/breadth 1.5	Preparation of slab as described in I..S.2* May be laid in long strips, and sawn into bays of 25 m² area
Unbonded	Not less than 75mm	Not more than 10m²	Some curling is likely even with limited areas. The use of 20mm maximum size coarse aggregate may help to limit curling. To reduce it further, over-slabbing 100 to 125mm thick may be used

Source: *Hard Landscape in Concrete* by M. Gage and M. Vandenberg, Architectural Press, London, 1975
*See source

Bonded construction is used when monolithic construction is not possible. In this case the topping is not included as part of the structural depth of the slab. The surface of the slab must be carefully prepared to insure that the topping is keyed into it. Due to the thickness of the topping there is a tendency for the bond to be broken when the topping shrinks in setting.

Unbonded construction is that which results when the structural slab is separated from the topping by a damp proof, insulating or isolating membrane. This is sometimes used when it would not be economical to prepare the surface slab for bonding. However unbonded construction is the least satisfactory of the three methods and it is difficult to avoid curling of the topping. In this case also, of course, the topping cannot be considered as part of the structural slab.

Scabbled finish An exposed aggregate finish can be produced on the surface of a slab which has already hardened by the use of a scabbling machine. These machines are simple to operate and can be particularly useful for upgrading the surface of an existing pavement.

SELF DRAINING CONCRETE PAVING
This pavement is constructed of no-fines concrete which is water permeable. The surface can therefore be constructed without a fall. The concrete pavement consists of two courses. The top course is approximately 105mm thick and comprises two layers.

The upper wearing layer is 40mm thick and is of no-fines concrete, using small size aggregate. It is constructed monolithically on top of a 65mm thick layer using medium size aggregate. Below the top course the 100mm sub-base may be composed of dry material, lean concrete or large size aggregate, no-fines concrete.

9.28

chevron

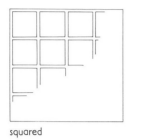
squared

hobnail

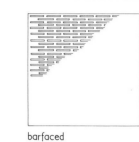
fluted

barfaced

9.29

The water which drains through the top course is collected in a waterproof membrane and carried to land drains. The membrane is situated either between the two courses or below the sub-base. It is given a fall of at least 1 in 80.

At the present time this material is still in the experimental stage and it is being developed for use in pedestrian and play areas. It is mentioned here because of its apparent potential. If it could be developed into a viable type of pavement for parking areas and lightly trafficked roads it would be a valuable tool in the hands of the designer.

There are however certain problems which must first be overcome. One is that of repair after the pavement has been disrupted. Also, if the pores of the concrete become silted up, rain water will lie on the surface, and the long term effects of frost in wet conditions are not yet known.

Construction: A sub-base of dry material will consist of well-compacted blinded hardcore or granular material (for example hoggin) constructed to falls with a minimum thickness of 100mm. If a lean concrete sub-base is used it should have an aggregate/cement ratio of 18 and a water/cement ratio of 0.42 to 0.50. The maximum size of aggregate should be 40mm. These sub-bases should be covered with a waterproof membrane consisting of a 500 gauge polythene sheet.

If a no fines concrete sub-base is used it should be constructed on top of a 500 gauge polythene sheet laid on formation. The thickness of the no-fines concrete will be 100m with an aggregate/cement ratio of 6, and a water/cement ratio of approximately 0.42, the exact value being found by trial and error.

The lower layer of the top course will have approximately 65mm minimum compacted thickness using 10mm single size aggregate.

The top layer of 40mm compacted thickness will have 5mm to 2.3mm, non-standard, single size aggregate. The top layer should be constructed within two hours of the lower layer but both should be compacted separately. The aggregate/cement ratio for both layers should be 4:1. It will probably be found that the water/cement ratio will be rather higher than in the no-fines sub-base and may vary between 0.45 and 0.55. During the construction period care should be taken to ensure that the porous texture of the no-fines concrete is not destroyed. In particular water content and compaction are critical features.

UNIT PAVING

This type of paving consists of precast elements, laid side by side to form the road surface. It can be used in conjunction with a flexible or rigid pavement construction.

Precast concrete paving slabs

Precast slabs are most frequently used for pedestrian areas. However, a choice of thickness is available and when the thicker units are employed and adequately bedded they are suitable for surfaces which are used by light vehicular traffic, for example for car access to a small group of dwellings, **9.28**, or for short service vehicle access roads.

These units can be very effective in the integration of the road surface with surrounding built forms. A single size of unit can be laid or alternatively, pleasing effects can result from the judicious use of more than one size. Also the creation of pattern and alterations in scale may be achieved by using the precast units in association with other materials, for example brick, granite setts or interlocking concrete paving blocks.

The capacity of the thicker slabs to support wheel loads is a function of the quality of the sub-grade material and of the pavement construction. Since the manufacturing process cannot usually be adapted to the inclusion of reinforcement, extra load-bearing capacity is obtained by an adequately designed sub-base, slab thickness and a

restriction on the slab surface dimensions to say 250–300mm.

The normal thicknesses manufactured are 50mm for pedestrian traffic and 63mm for occasional vehicular traffic. Slabs 75mm and 100mm thick are also obtainable. The standard plan dimensions are listed in Table 6. In addition to square and rectangular units other shapes are available including hexagonal, wedge shaped and circular.

TABLE 6 Standard sizes for precast concrete slabs

Type A	600mm × 450mm
Type B	600mm × 600mm
Type C	600mm × 750mm
Type D	600mm × 900mm

In all cases, thickness 50mm for pedestrian use; 63mm for occasional vehicular traffic.

In the manufacturing process the depth of the slab is more easily altered than the shape and so non-standard shapes are likely to be expensive. The maximum size procurable is limited by the capacity of the manufacturing machinery.

The British Standard relating to paving flags is BS 368:1971, Precast concrete flags—metric units. Standard flags are normally manufactured with Portland cement to BS 12 or Portland blast furnace cement to BS 146. Aggregates comply with BS 882:1201 or alternatively with BS 1047: Part 2 1974, Air-cooled blast furnace slag coarse aggregate for concrete. A number of different manufacturing methods are used and various textures and types of finish are available, some of which are outlined below. (Important reference works on this topic are *Hard Landscape in Concrete* by Gage and Vandenberg[40] and *Guide to Exposed Concrete Finishes* by Gage.[41])

MANUFACTURING The principal methods of manufacture are the hydraulically pressed, vacuum pressed, dry-pressed-vibrated, and dry-pressed-tamped processes. These manufacturing methods employ low water content of the cast slab to obtain quick strength and high quality. The quality of slabs with respect to strength, durability and frost resistance is laid down in BS 368:1971, Precast concrete flags.

In the hydraulically pressed method the slabs are formed in mould trays, the bottoms of which are covered with a fine wire mesh. On top of the mesh is placed a sheet of paper to prevent the fines of the concrete being forced out. A wet mix is used and the water is dispelled through the paper membrane by the application of hydraulic pressure. It is possible to vary the texture by placing assorted types of wire mesh in the mould. The normal finish is a smooth, slightly pimpled surface.

This is the method most commonly used and if de-icing salt is likely to be spread it is essential that hydraulically pressed slabs are employed.

The vacuum press process involves the application of pressure to a wet mix and surplus water is extracted by means of a vacuum pump. The finish is similar to the hydraulically pressed one but a rough texture is also available.

In both the dry-pressed-vibrated and the dry-pressed-tamped process an 'earth-dry' mix is used. In the former process, pressure and vibration are employed and in the latter the units are compacted by mechanical tamping action. Both dry pressed methods can be used to manufacture slabs which are smooth, surface textured or with exposed aggregate. With the vibrated slab process special textures can be obtained by casting the mould face down. But rough textured or aggregate faced finishes are usually formed on the upper surface.

Precast units are available in a wide range of colours although there is a tendency for too much colouring additive to be incorporated. A more subtle use of colour in the precast units is likely to be the most effective. The colouring agent may be incorporated only in the top 12mm or it may be used throughout. Pigments should be in accordance with BS 1014:1975.

The variety of textures obtainable with hydraulically pressed slabs is indicated by **9.29**.

For the economic mass production of precast units in a factory the choice of surface finishes tend to be limited by the manufacturing process, although with the vibrated slab method the surface can be given any finish which is available for the normal construction of a precast unit in a mould. These can be categorized under the headings 'face down finishes' and 'face up finishes'.

Face down finishes These are applied to the bottom of the mould before casting and are only exposed when the unit is removed from the mould and turned over to present the finished surface. In this method profiled or textured bottom liners can be made with timber, plastic, concrete, metal or rubber; consequently many different surface textures are possible. Other methods exist for obtaining a special finish with a face down technique. A retarder can be applied and the aggregate exposed by brushing and spraying after removal from the mould or the more expensive expedient of tooling the surface after the concrete has set can be adopted.

Although the brush and spray technique does not produce such consistent results it has the advantage of being applicable as soon as the unit is demoulded whereas a three to four week period must elapse before the hardened surface may be tooled. This can be a serious drawback when a large number of units are to be produced. The difficulty can be overcome by a further method, abrasive blasting, in which the hardened cement paste is removed from the surface by a jet of grit, sand or shot and compressed air. This process, which produces a uniform colour and texture, can be conveniently integrated into a mass production process.

Another face down finish is obtained by the aggregate transfer process. The aggregate is placed by hand on a sand bed on the bottom of the mould; a layer of sand/cement mortar is then placed over the aggregate and finally the concrete mix is placed on top. After removal from the mould, the loose sand is brushed from the surface. This is an expensive, labour intensive procedure but excellent results in terms of appearance can be obtained. It may be appropriate to consider its use in special circumstances for small areas when a very high quality finish is required. The appearance of the finish will depend on the thickness and colour of the sand bed as well as the colour, shape and size of aggregate. The aggregate size used is normally from 40mm upwards. This is a technique which is sometimes used in the manufacture of precast panels for the external walls of buildings. If proposed for a running surface, test

pancls may be required to ensure adequate embedment of the stones.

Face up finishes These are produced when the finished surface is that exposed on the top of the casting in the mould. The face of the unit can therefore be worked either before or after demoulding and all the finishes are available which are used for in-situ concrete including rolling, floating, tamping or sawing the surface before the concrete has set, brushing and spraying after initial set or tooling after the concrete has hardened.

A further option is that of a broadcast finish in which a layer of small aggregate, carefully placed on the surface of the wet concrete, is partially tamped in to obtain an exposed aggregate finish. After embedment the surface is cleaned by spraying with water. Some structural mixes are suitable for this process but with others it is necessary to use a special topping mix.

CONSTRUCTION Precast concrete units should be laid on a complete, dry mortar bed consisting of 1 part cement or lime and 3 parts sand by volume. The mortar bed should be spread evenly over the sub-base to a nominal thickness of 25mm, the actual thickness being between 12mm and 37mm thick.

The five spot method of laying must not be used.

The slabs should be butt jointed with a dry mortar mix of 1 part cement to 3 parts sand by volume and any surplus material removed. The laying technique is detailed in the Cement and Concrete Association booklet *Laying Precast Concrete Paving Flags*.[42] With regard to vehicular access to paved surfaces this publication states: 'Where it is likely, or intended, that vehicles up to light commercial vehicle rating will mount or cross the paving, flags 63mm ($2\frac{1}{2}$in) thick should be used, laid and bedded on a concrete sub-base not less than 75mm (3in) thick. The proportions by volume of the sub-base concrete should be 1:2:4 of cement: fine aggregate: coarse aggregate. For vehicular crossings that are to be used by heavier commercial traffic, the thickness of the concrete sub-base should be increased to 150mm (6in) or alternatively, the crossing should be constructed entirely in in-situ air-entrained concrete. The proportions by volume of this concrete should be 1:2:3 of cement: fine aggregate: coarse aggregate and the slab thickness should be 150mm (6in)'.

For heavier loading consideration should be given to the use of 75mm or 100mm thick flags on an adequate sub-base. The total pavement construction should be designed in accordance with good practice in the light of the proposed loading and sub-grade conditions.

Some possible laying patterns are shown in **9.30**.

Hollow concrete paving with grass

The large areas of hard surfacing for vehicle access and parking to be found in residential areas are often intimidating and out of scale with their surroundings. This is particularly the case when parking strips are supplied at the roadside. On the other hand if vehicles are allowed to park on grass areas the grass is quickly killed and in wet weather the surface cannot be used. The incorporation of grass/concrete paving for hard standings can be a solution to this problem, especially when the hard surface is adjacent to an area which is grassed in the normal manner. In some cases it is possible to completely avoid the breaking

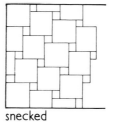
snecked

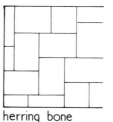
basket weave

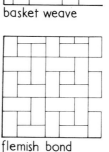
herring bone

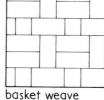
flemish bond

9.30

9.31

up of a soft landscaped area by an asphalt or concrete road. If the proposed road is for use as a fire path for example, it is often feasible to design a sufficiently strong and durable running surface with the use of grass/concrete paving.

An example of the use of this type of surface to provide parking space in residential areas is shown in **9.31**.

Grass/concrete surfaces can be constructed in urban or rural settings. They can be adopted to provide parking areas in front of or behind dwellings and in the countryside they may be useful in providing hard standings to allow motorists to park at vantage points off the road to enjoy areas of special scenic beauty.

This surface material consists of a concrete grid, the hollow parts being filled with soil in which grass is grown. The grid

may be constructed of precast units or in-situ concrete and may be single or multi-cellular. Units are designed to support the wheel loads without damaging the grass and to have sufficient area to adequately distribute the loads to the formation level.

The grass tends to overgrow the concrete so that the appearance of the surface becomes that of a grassed area. Depending on the design and the system used the pavement will be able to support light everyday traffic and occasional heavy loads.

The mowing procedure for this surface is the same as for ordinary grass and it is self-draining, provided the subsoil is of ordinary permeability. The concrete grid is normally constructed on a bedding layer of sharp sand which is laid on a consolidated sub-base. In certain conditions it may be possible to dispense with the sub-base. For example it may not be required when constructing the paving on a sandy or gravel subsoil. The pavement should always be constructed in accordance with sound engineering principles and in consultation with the manufacturer. Filling and, seeding should be carried out in the light of the manufacturer's recommendations.

MONOSLABS *(Manufactured by Mono Concrete Ltd)*
This surface consists of precast concrete grid units containing soil chambers and having regularly spaced concrete projections for wheel contact at the surface. The surface area comprises 75% grass and 25% concrete whilst the bearing area consists of 88% concrete. There is one basic unit, measuring 600mm × 400mm × 125mm which can easily be handled by two men, **9.32**.

In ground of normal permeability no special drainage is required since the concrete grid allows all the rain water to soak through. Freshly laid and sown Monoslabs should be treated as a normal area of grass.

Construction The sub-base design will depend on wheel loads and prevailing ground conditions. The manufacturer's recommendations for average conditions are shown in **9.33**.

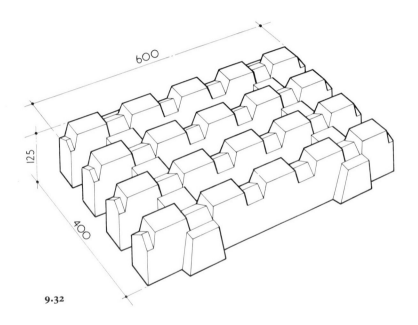

9.32

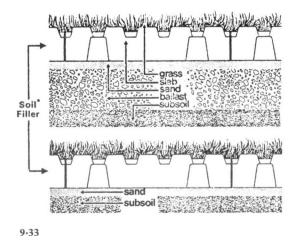

9.33

For heavy vehicles
For roads carrying heavy vehicles a 100–150mm layer of quarry waste or similar is required as a sub-grade. This should be quite stoney—up to 75mm—but containing fine material, soil or loam from which grass roots can find food. This material must *not* be washed and should be followed by a bedding layer of sharp sand 20mm thick.

For light vehicles
Providing the soil is permeable and the ground level, a layer of well tamped sharp sand to a depth of 20mm is all that is required as a foundation.

241

The slabs are laid in regular rows without jointing and firmed into place, **9.34**. Clean friable soil or soil with a peat mixture should be levelled off 30mm below the top surface. The grass seed is then sown and covered with a thin layer of fine soil and levelled with a hard broom. The level will settle to about 20mm below the top surface and this is important for germination. Seed should be sown while soil is still loose after filling. It is recommended that the grass be cut with a 'flymo' type mower.

Theory The theory behind the design of the surface configuration of monoslabs is that the wheel loads are taken by the concrete projections and distributed, thus protecting the soil from compaction. Consequently the grass is provided with good conditions for germination and growth with the result that it can thrive under what would otherwise be adverse conditions, including the use of the surface for parking cars.

9.34

GRASSCRETE *(Manufactured by Grass Concrete Ltd)*
Grasscrete is an in-situ process the purpose of which is to combine the general appearance of natural grass with the ability to support vehicle loads without compaction of the soil or detrimental effects on the grass. The aim is to combine the pleasing appearance of a grass surface with the strength and durability of continuous reinforced concrete. Permanent plastic formers are used, **9.35**, to provide full depth, cruciform, tapered voids at regular spacing thus allowing natural drainage, where the sub-soil conditions are suitable. After the concrete is set the tops of the formers are burnt out and the voids are filled with soil and seeded. Two types of plastic formers, manufactured from 0.75mm rigid polystyrene sheet, are available, **9.35**.

GC1 FORMER
600mm × 600mm (2' × 2'), 100mm deep (4")
Formers per sq. metre 2.78
Concrete   1 cubic metre = 16 sq metres
Soil   1 cubic metre = 18 sq. metres

GC2 FORMER
600mm × 600mm (2' × 2'), 150mm deep (6")
Formers per sq. metre 2.78
Concrete   1 cubic metre = 12 sq. metres
Soil   1 cubic metre = 12 sq. metres

The square mesh fabric reinforcement, to BS 4483, fulfils the functions of positioning formers, controlling surface cracking and providing tensile strength to offset any uneven settlement of the sub-base. The manufacturers recommend the use of a concrete with a minimum 28 day strength of $28N/mm^2$, using maximum aggregate size of 10mm, a minimum slump of 100mm $\pm$ 18mm, and 3% $\pm$ 1% air entrainment.

The thickness and composition of sub-base, depth of slab and amount of reinforcement will depend on subsoil conditions and vehicle loadings.

Construction The construction sequence is indicated in **9.36**.

HEXPOTS *(Manufactured by Mono Concrete Ltd)*
Hexpots are precast hexagonal units with a hollow core. They have been designed to provide permanent, decorative hard standings and approach routes over grass without impairing the turfed aspect where wheeled traffic is only occasional.

The units are manufactured using high grade, low workability Portland cement concrete vibrated in steel moulds. They are made in two sizes, **9.37**

Construction The manufacturers specify that Hexpots should be laid using a 20mm thick layer of ash or sand blinding on at least 150mm of hard core and that they should be bedded and jointed in cement mortar. A typical section is shown in **9.38**.

GRASSCEL *(Manufactured by Brooklyns Westbrick Ltd)*
This material provides a grass/concrete finish with applications which include the construction of surfaces subject to light vehicular traffic; car parks, hard standings and access. Its uses are similar to those of Monoslabs and Grasscrete.

This is a perforated precast concrete slab, moulded on its upper face so that the finished surface is 75% grass and 25% concrete. There is one basic unit, **9.39**, based on a modular grid of 600 × 400mm. The unit is unreinforced and the concrete has a 28 day cube strength of 35 N/mm^2.

Construction A porous flexible sub-base is constructed. The material used and the depth of the sub-base will depend on the frequency and intensity of loading and the local ground conditions. If the existing soil is stable and the traffic light it may be possible to dispense with a sub-base. A laying course formed from sharp sand is placed on the sub-base, screeded and levelled to a thickness of 20mm, and lightly compacted.

The dry-jointed Grasscel units are placed edge to edge and consolidated in position using a wooden tamping board.

The slab voids and channels are filled with a good quality top-soil to within 10mm of the top of the slab and the grass is sown. A further 10mm of fine soil is then added and levelled.

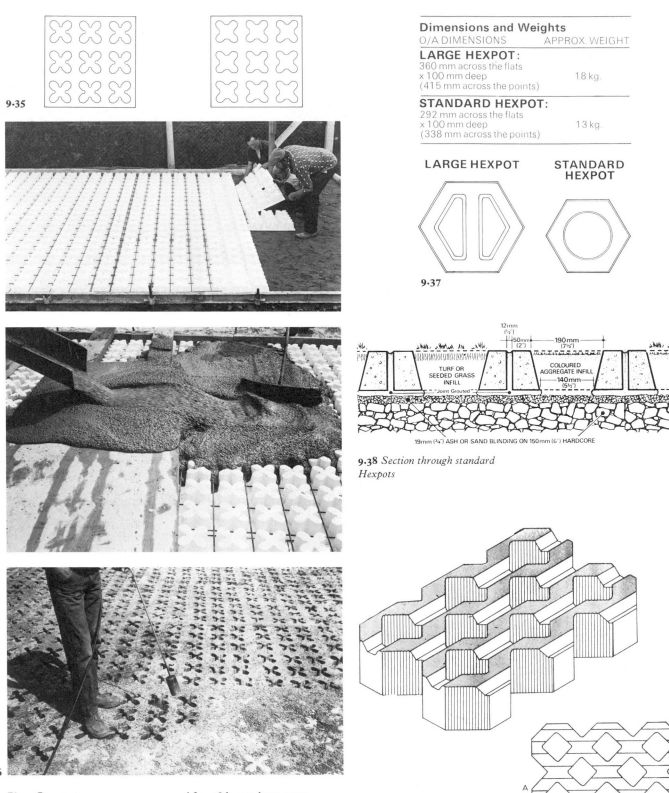

9·35

Dimensions and Weights

O/A DIMENSIONS	APPROX. WEIGHT
LARGE HEXPOT: 360 mm across the flats x 100 mm deep (415 mm across the points)	18 kg.
STANDARD HEXPOT: 292 mm across the flats x 100 mm deep (338 mm across the points)	13 kg.

LARGE HEXPOT   **STANDARD HEXPOT**

9·37

12mm (½")
50mm (2")
190mm (7½")

TURF OR SEEDED GRASS INFILL

COLOURED AGGREGATE INFILL 140mm (5½")

Joint Grouted

19mm (¾") ASH OR SAND BLINDING ON 150mm (6") HARDCORE

9·38 *Section through standard Hexpots*

Plan

600

Side elevation

400

120

End elevation

9·39

·36

Place Grasscrete polystyrene formers on a well compacted sub-base with 10–20mm sand blind used for levelling

Reinforcing steel mesh is laid in position on preformed spaces provided within each former

Pour concrete onto boards before spreading to avoid damage to the formers. Screed level with top of formers. After initial set, brush to leave the tops of the formers exposed

After 48 hours, burn away exposed former tops

Fill voids with good quality weed-free top soil

After settlement, seed and top up with soil using the correct grass for the contract

Fertilize as for normal grass surfaces

9.40

9.41

9.42

9.43

Settlement due to watering or rain should be topped up. The sown grass should be treated in the same way as any other grass sown area. Mowing cuts should be made diagonally across the units.

Concrete block paving
Concrete block paving has many advantages over the more common methods of surfacing roads, especially in residential and other areas where the townscape qualities of the environment are important. The scale of the blocks make them an ideal surfacing material for access roads and dual use areas. The small scale and distinct texture of the paving give it a visual appeal which is lacking in large areas of plain concrete or asphalt. Moreover the fact that it has obviously been laid by hand gives it a cared-for look which is imparted to the space as a whole. This type of paving is particularly suitable in areas where slow vehicle speeds are desirable since the small scale and high quality finish are a psychological deterent to high speed and in addition the texture of the surface is likely to dissuade drivers from moving excessively fast. It is also commonly used in areas subject to heavy slow-moving traffic such as docks and container terminals.

Differences in texture, colour, shape and bond can all be utilized by the designer both to embellish the surface and indicate usage. Markings denoting parking bays and pedestrian crossings, lines between lanes, areas denoting special usage and even lettering can all be made using different colours of blocks. Such lines and areas are easily removed when no longer required by replacing with blocks of the basic overall colour.

A further attraction of the material is its link with the past since the old vernacular road surfacings comprised small units: bricks, cobbles, setts and wood block paving.

Construction with concrete paving blocks has been developed in Europe in the last twenty-five years, especially in Germany and the Netherlands, and is now being adopted to an ever increasing extent in the UK.

Whether the blocks used are the plain rectangular type or those with an indented profile, the herringbone pattern when feasible will give the best resistance against horizontal movement. In the case of rectangular blocks it is essential to use the herringbone pattern when designing for vehicular traffic. With the indented profile blocks, stretcher bond can be used. The blocks are light to handle and easy to lay, **9.40**. An experienced three man team can pave up to 150m² of surface in a day. When properly laid and bedded on an adequate sub-base the surface will support the wheel loads of heavy lorries without horizontal or vertical movement of the blocks. After laying, portions of the road surface are easily lifted to obtain access to services and such areas can be simply reinstated without leaving any visible trace of the disturbance. This is in marked contrast to the reinstatement of asphaltic or in-situ concrete pavements when visible patches invariably result.

Illustration **9.41** depicts typical shapes of blocks currently available in this country. Most can be obtained in natural grey, buff, red, dark grey and black. White rectangular units are also available and the rectangular blocks can be obtained with conventional square edging as well as a chamfered edge.

It is important that the very attractive surface of concrete block paving is not impaired during construction. In this period the road surface, of a residential area especially, is at greater risk of disfigurement due to split materials or of damage due to excessive loads from heavy equipment or the storage of materials, than at any later time. Consequently there is an advantage in postponing the laying of the blocks until the building operations are substantially complete. A normal sub-base can be designed to support the construction traffic or a 75mm thick concrete slab can be used as described under the heading 'Sub-base' below.

Concrete block paving can be equally well used for curved as for straight roads, **9.42** and **9.43**. Very rich and pleasing effects can be produced by using, in a meaningful way, different shapes of precast units in association with each other, **9.44**.

Detailed information relating to this material is contained in the following Cement and Concrete Association publications: *Laying Concrete Block Paving,*[13] *Concrete Block Paving for Lightly Trafficked Roads and Paved Areas,*[14] *Concrete Block Paving for Heavily Trafficked Roads and Paved Areas*[15] and *A Specification for Concrete Paving Blocks.*

9.44

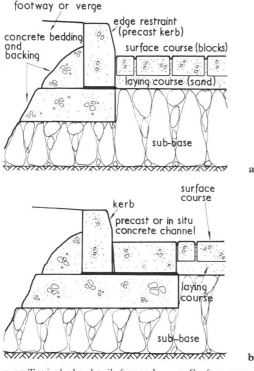

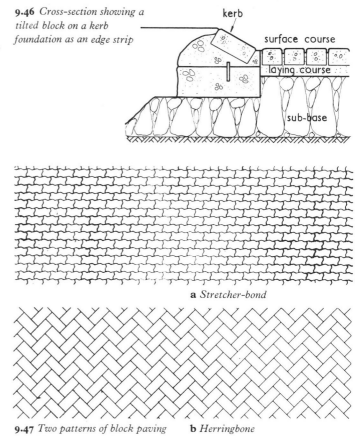

9.45 *Typical edge details for roads with concrete block paving. The kerb is shown laid on a foundation within the sub-base. If a concrete sub-base is used, the kerb is bedded directly in it*

a *Surface course abutting kerb*
b *Surface course abutting concrete channel*

9.46 *Cross-section showing a tilted block on a kerb foundation as an edge strip*

9.47 *Two patterns of block paving suitable for residential roads*

a *Stretcher-bond*
b *Herringbone*

DESIGN The pavement for a concrete block road consists of a sub-base, laying course, the layer of interlocking precast blocks and edge restraint to prevent the blocks from spreading. For heavily trafficked roads a roadbase is used. (These notes on design and construction are based on the Cement and Concrete Association publication for lightly trafficked roads.[44]) The four elements are shown in **9.45a**, **9.45b** and **9.46**. The edge detail with tilted blocks is normally considered for lightly trafficked roads rather than those taking large volumes of vehicles.

The sub-base The blocks can be laid on top of a sub-base designed in accordance with the recommendations given in the Transport and Road Research Laboratory's Road Note 29, the blocks taking the place of a normal surfacing and roadbase.

In *The Design of Concrete Block Roads* (Cement and Concrete Association, 1976) Knapton showed that the blocks have a load spreading capability similar to that of 160mm of rolled asphalt. If this is taken as a basis of design it is possible to use Road Note 29 for the determination of sub-base thickness for any subgrade and pavement life expectancy for roads carrying up to 1.5 million standard axles.

Type 1 and type 2 roads, as defined in the Road Note, are

TABLE 7 Flexible sub-base thicknesses (mm) for various sub-grades and types of road when the water table is more than 600 mm below formation level [1]

Type of road	Design life (years)[2]	Heavy clay	Silt	Silty clay	Sandy clay	Well graded sand or sandy gravel
				Type of sub-grade		
1 Cul-de-sac or other minor residential road	40	400 (550)	400 (550)	190 (300)	140 (230)	80 (80)
2 Through road or road carrying regular bus routes with up to 25 public service vehicles per day in each direction	40	450 (600)	450 (600)	220 (340)	170 (260)	150 (150)
3 Major through road carrying regular bus routes with 25 to 50 public service vehicles per day in each direction[3]	20	440 (590)	440 (590)	210 (340)	160 (260)	150 (150)

Notes:
1 The figures in parentheses should be used if the water table is less than 600 mm below formation level
2 If other design lives are to be considered, direct reference should be made to Road Note 29
3 Road Note 29 must be consulted if the sub-grade is frost-susceptible

normally designed for a life of 40 years, and type 3 for 20 years. The thickness of flexible sub-base required for five types of sub-grade is given in Table 7, using the recommended design life. This table is reproduced from the Cement and Concrete Association publication *Concrete Block Paving for Lightly Trafficked Roads and Paved Areas*, and is based on Road Note 29. The method of sub-base construction is as described in the Department of Transport's *Specification for Road and Bridge Works*. Suitable sub-base materials include those detailed in clauses 803, 805, 806, 807 and 815 of that publication.

An alternative form of sub-base is a thin concrete slab, designed to support site traffic during construction and thus ensuring that the paving is not at risk during this period. Research carried out in the United States[46] tested a range of slabs with and without sub-base on sub-grades with a California Bearing Ratio (CBR of 2%) and upwards. The results indicate that an unreinforced slab 75 mm thick laid directly on a sub-grade with a CBR of 2% or more will carry up to 0.3 million standard axle loads of 8.2 tonnes before failure. This is a much higher load than would be imposed on a construction site road serving 1000 houses in a residential development, so it can be assumed that such a slab is adequate as a haul road and subsequently as a sub-base for block paving in a road serving up to this number of dwellings.

The slab does not require joints or reinforcement but should not be thinner than 75 mm at any point and the surface tolerance should be within ± 20 mm.

Maximum aggregate size should be 20 mm and the concrete should have high workability with a 28 day characteristic cube strength of 30N/mm^2 if site traffic is kept off the surface for 14 days and 40N/mm^2 if site traffic is allowed on after only two or three days. In the latter case the ambient temperature should remain above 10°C between the time of laying and the opening to traffic.

Laying course The laying course of washed sharp sand should be spread to a profile similar to that of the surface and so that it will give a compacted thickness of 50 mm. It should contain not more than 3% of silt and clay by weight, and have not more than 10% retained on a 5 mm sieve.

Surface course The paving blocks are closely fitted with the joints subsequently filled with sand. The blocks must interlock to give proper load distribution. This interlocking is produced by the use of proprietary shaped blocks or by laying rectangular blocks in a herringbone pattern, **9.47**. It is unnecessary to use blocks more than 80 mm thick for residential roads. For type 1 roads, drive entrances and other areas carrying very light traffic, blocks not less than 60 mm thick may be used.

Edge restraint The tight joints and the interlocking action between the blocks enable vertical loads to spread through transverse distribution and horizontal loads, due to acceleration and deceleration, to be distributed through the surface. Because of the importance of the interaction between the blocks it is necessary to provide restraint at the edges to prevent spreading. This can be accomplished by a precast or in-situ kerb or other suitable unit which may be flush or upstanding. Some manufacturers of blocks will supply special edge filling blocks. **9.45** shows two possible uses of upstanding precast kerbs which fulfil this restraining function.

CONSTRUCTION The paving blocks will preferably not be laid until the other building operations on the site are substantially completed so, before laying the blocks, the sub-base should be cleaned and repaired where necessary. The bedding sand should be screeded to an unconsolidated thickness of about 15 mm in excess of the final consolidated thickness of 50 mm. The surface is not vibrated until after the paving is laid so the unconsolidated depth of sand should be determined by trial at the commencement of laying and checked periodically as work proceeds. The sand is screeded after spreading using the kerbs as guides if the road is sufficiently narrow, otherwise temporary screed rails are used.

There are three stages in the construction of the surface course: laying the blocks, completing the edges with cut blocks where necessary, and vibrating the paving. The first few rows should be placed with particular care then the block layer can work more quickly, placing each subsequent row hard against the paving already laid. It is necessary to ensure at all times that a tight fit between blocks is being attained. The laying of cut blocks at edges, around manhole covers and so on is left until later. The laying of blocks is a simple and rapid operation and it is important than an efficient routine is set up to keep the block layer continuously supplied with blocks within easy reach. Non-standard shapes are filled with portions of standard units cut with a block splitter or by sawing.

When the placing of cut blocks is completed the paving is consolidated by the use of a plate vibrator. Normally two or three passes of the vibrator will be adequate but the actual number of passes should be determined by trial. The blocks should be vibrated sufficiently to ensure that no further consolidation from traffic loads will occur and that a smooth surface has been established.

Sand is then brushed over the surface and two or three more passess made with the vibrator to establish the sand in the joints and complete the interlocking of the units. The road can then be opened to traffic.

SETTS

Granite setts, for a long time a traditional road surfacing material, are now very expensive although they can sometimes be obtained more cheaply from local authorities depending on the extent and location of redevelopment work. The smoother driving qualities of asphaltic pavements led to the replacement of many street surfaces in which setts had been used. However now that the attractive appearance of setts and the desirability of slow vehicle speeds in the urban environment are better appreciated the remaining traditional surfaces are less likely to be replaced. The granite setts for re-use which come onto the market vary in size from 4 ins cubes to 12 ins × 6 ins × 5 ins deep. They should be laid on a correctly designed sub-base which has been blinded with ash. A 13 mm bedding layer of sand is prepared to receive the setts which are placed in the adopted pattern leaving 10 mm joints with the smaller units. The joint size may be increased with the larger setts if the designer so wishes. Small chippings are brushed and rammed into the joints which are then sealed with a material appropriate to the proposed usage. This may be cement grout or mortar, hot or cold bituminous materials, sand or soil.

The more complex traditional patterns of laying are very attractive but the most economic pattern is likely to be that of the common brick stretcher bond.

Sandstone setts can also be obtained and perform satisfactorily in spite of being less hard than the more usual granite.

Concrete setts are not uncommon in Belgium and have been used for many purposes including the paving of pedestrian streets in Liege, Brussels and Louvain-La-Neuve. In an article entitled 'Concrete setts from Belgium' in *Concrete Quarterly*,[47] the setts shown in **9.48** and **9.49** were illustrated. The article reports that these 'are known as Beton Blanc de Bierges and are made by SA Anciens Etablishments Jean Delvaux of Bierges-Wavre in Belgium. They were developed in 1972 by the manufacturers and a design team of the Groupe Urbanisme et Architecture under the leadership of Prof. R. Lemaire of Louvain University. The result is a series of modular units which received a special award for their use at Louvain-La-Neuve (Signe d'Or 1974 de l'Industrial Design)'.

In the development of these blocks colour and texture were studied particularly when the surface is wet so that a warm appearance would result even in wet weather.

140mm and 70mm square setts are manufactured and also a 210 × 140mm unit for use with a bonded pattern. 80mm and 70mm depths are used for vehicular traffic and 40mm for pedestrian areas. The following description of the pavement construction is reproduced from the article referred to above:

'Where 80mm thick setts are used, a 200mm bed of ordinary sand is first laid on good consolidated earth. A layer of stabilized sand 120mm thick is then laid, on top of which the setts are placed. The surface is then vibrated and brushed with stabilised sand to fill the joints. The thinner 40mm setts are laid on a 200mm bed of lean concrete and mortar. Kerbs and edge units are incorporated in the design of the paving.' The setts have a compressive strength of 600kg/cm^2.

In this country the use of precast concrete rectangular and indented paving blocks is becoming increasingly more common but the use of concrete setts as a paving unit is not yet considered an attractive option. The firm of Mono Concrete Limited manufactures a precast concrete paving

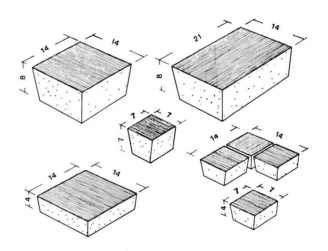

9.48 *Different types of concrete setts (dimensions in centimetres)*

9.49

248

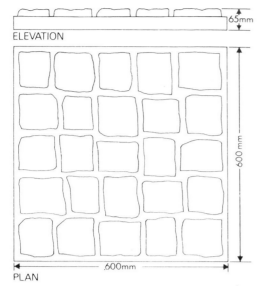

ELEVATION

PLAN

9.50 *Thaxted concrete paving slab manufactured by Mono Concrete Limited (weight 44 kg)*

slab under the name of Thaxted paving, **9.50**. The material has the appearance of granite setts and the firm recommends it for use as decorative paving and as deterrent paving to indicate areas of restricted pedestrian movement. *Brick setts,* in engineering quality conforming to BS 3921, are manufactured by Barnett and Beddows Limited of Staffordshire. The basic standard size is a 100mm cube and there are some standard special shapes. The units can be supplied to a variety of dimensions for special purpose if required. All can be supplied with a wirecut or smooth textured finish.

COBBLES

Cobbles are usually water-rounded stones obtained from river beds or beaches. But they can also be procured from other sources including gravel pits, where flint rejects can be inspected, cement works using chalk rather than limestone as a raw material and potteries where wash mill flints can sometimes be obtained. Cobbles are sold graded in size and those used for paving are usually in the range 36mm to 100mm. They are suitable for deterrent paving and can also be used on a running surface for short lengths when the vehicular traffic is light; say on a short driveway or in a courtyard. The colour of natural stones available includes white, grey and brown. Flint cobbles are grey in colour and have a pitted surface. The wash mill flints from the potteries are also grey and pitted, resembling pumice in appearance.

Three patterns for laying cobbles are indicated in **9.51**.

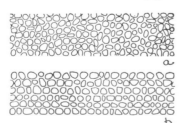

9.51 *Patterns for laying cobbles : (a) random pattern (b) course pattern and (c) flat cobbles laid parallel*

Linear patterns are more pronounced when elongated rather than the more spherical stones are selected.

Construction Cobbles are usually set in concrete though a traditional method of laying was to embed them by ramming in compacted soil, sand, fine gravel or a mixture of clay and gravel on a hard foundation. When they are set in concrete they will look most satisfactory placed close together so that the concrete base is inconspicuous.

Typical of the type of detail used for a cobbled surface is that shown in **9.52**. The thickness of the concrete and hardcore will be a function of the traffic loading and the subgrade conditions. For light vehicular traffic a minimum of 100mm of concrete would be required.

STONE

Stone is a relatively expensive paving material and is normally used in small areas intended primarily for pedestrians. Nevertheless with adequate pavement construction and appropriate bedding vehicular traffic can be supported.

There are a number of stones sufficiently hard to be utilized for outdoor paving and with their slight irregularities and natural weathering characteristics they make attractive surfaces. The riven finish (split along the bedding planes) and axed finish are best for an informal surface and sawn or rubbed finish for a more formal effect.

Second hand material can sometimes be obtained from local authorities, especially York stone from old walkways. Second hand stone is cheaper to buy although it is more expensive to lay.

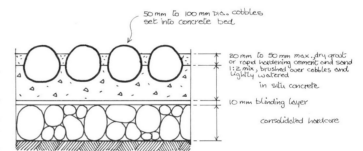

9.52 *Cobbles—a typical detail*

York stone This sandstone is the most common stone used for paving slabs. The surface can be riven, sawn or rubbed. The cheapest finish is riven which produces a pleasant rough texture. The colour ranges from light buff to brown and blues are also obtainable. Sandstones of similar appearance are quarried in Lancashire and the Forest of Dean. The thicknesses available range from 50mm to 63mm and the sizes go up to 900mm × 600mm. *Portland stone* This is an oolitic limestone and, being softer than York stone, it should only be used if the traffic is not likely to be heavy. It is sawn and can be used with the saw marks left on. The colour lies within the white to grey range and the thicknesses are in the range 50mm to 63mm. The plan dimensions available are similar to those for York stone. *Purbeck stone* from an adjacent limestone bed is harder. It is also more expensive.

Bath stone Since it is not hard wearing this stone should only be used where traffic is light.

Slate Slate is expensive but very hard and resistant to weathering. It should be used in its riven or sawn condition

otherwise it is liable to become slippery when wet. In any case, the sanded and rubbed finishes are more expensive. Thicknesses vary from 13mm to 50mm depending on the plan dimensions. Sizes range from 450mm × 230mm at 12mm thickness (riven) to 900mm × 600mm at 25mm thickness (sawn or rubbed). Larger sizes are obtainable up to 50mm thick. Those in the region of 450mm × 230mm are usually selected since they are less expensive than the larger sizes. Blue-greys and greens, some with brown streaks, are obtained from Lancashire, Westmorland and Cornwall and purplish and grey-black from Wales.

Granite Like slate, granite is very hard and resistant to weathering but also expensive. It is not usually found in large slabs. The single axed surface is best for pavements. Colours include greys, pinks and dark greens.

Quartzite Hardwearing and expensive, this stone is imported from Norway. The surface is usually riven. Colours range from silver-grey to grey-green and dimensions range from 600mm × 600mm × 12mm thick to 1500mm × 1200mm × 50mm.

Cast stone The manufacturers can produce reconstituted flags simulating the natural material from almost any stone using stone aggregate and pigment. The product is cheaper than natural stone and more expensive than precast concrete flags. They may be regarded, apart from their colour and texture, as similar to precast concrete units and subject to the same strength requirements. They are not covered specifically by a British Standard Specification but BS 1217: 1975 Caststone lays down standards for the quality of products manufactured 'from cement and natural aggregate for use in a manner similar to and for the same purpose as natural stone'.

BRICK PAVING

Bricks have been used as a surface material for roads since the time of the Romans. They are more commonly employed at present in Europe, particularly in Holland and Belgium, than in the UK, although the considerable attractions of brick are gradually increasing its popularity. With proper selection they can be used as paving for heavy duty as well as lightly trafficked roads.

Suitable bricks are available in a great variety of colours and textures. Colours include reds, orange, yellow, ochre, browns, purple, blue and units with intermingled colours. Among the textures available are smooth, sand-faced, sand-creased, combed, stippled and dragged. Beside the wide choice of colour and texture other advantages are the small scale and sympathetic appearance of the surface, the ease with which sections can be lifted and relaid if necessary, the good weathering qualities of the material and low maintenance costs. There is also the attraction of a wide choice of laying patterns.

Durability is particularly important for bricks used for paving since wetting is more severe than with vertical surfaces. Also paving is likely to be wet for much longer periods. These facts together with the effect of night cooling due to loss of heat by radiation make it more susceptible to frost action than normal brick construction. Bricks are also subjected to intermittent point loads from vehicle wheels, but for normal traffic any damage in the form of splitting of the bricks is more likely to be caused by an unsatisfactory sub-base or laying technique than low

strength of the brick, provided it is one suitable for external use.

Traffic can break off the brick arrises if the pointing is too low. This is prevented by the use of flush or slightly dished joints. The sharp arrises of calcium silicate bricks are particularly vulnerable and for that reason are best not used where they will have to withstand very heavy wheel loads. The most important British Standards relating to brick-work are as follows:

BS 3921:1974, Specification for clay bricks and blocks: Reference is made to standard metric units and standards of size and quality are specified.

BS 187: Part 2: 1970, Specification for calcium silicate (sandlime and flintlime) bricks: This covers standard metric units, brick size and quality, and recommended mortar mixes for different uses.

BS 4729:1971, Metric and imperial units, Specification for shape and dimensions of special bricks: This relates to the commonly used, specially-shaped bricks for all purposes.

UNITS SUITABLE FOR PAVING

Clay bricks (BS Standard and metric modular) BS 3921 describes three classes of brick:
1 For internal walls
2 Facing and common bricks of ordinary quality
3 Facing and common bricks of special quality.

In addition eight grades of load-bearing brick are described and two grades of engineering brick, A and B.

The eight grades of load bearing brick and the two grades of engineering brick are designated according to strength; the engineering bricks have also an absorption specification. Since some bricks of low strength and high absorption can be very durable and are perfectly satisfactory for use as pavers the British Standard suggests that evidence of adequate performance over a period of three years is the best guide to a brick's suitability for paving. Otherwise special quality bricks having either a minimum strength of 48.5 N/mm^2 or a water absorption not greater than 7% should be used. The book *Hard Landscape in Brick* by Handyside[48] follows a statement of these BS suggestions with the recommendation that 'for earth-retaining walls, copings and paving, bricks should comply with the special quality requirements for soluble salts content and, whenever possible, should be checked for frost resistance by evidence of performance under conditions of exposure and of design details similar to the intended usage'. Later, considering resistance to exposure, he says that 'clay bricks and pavers should meet the requirements for bricks of special quality as defined in BS 3921', and when discussing strength, that 'the fact that most clay pavers fall within the engineering classes may be an indication that fairly high strength is needed if a wide range of conditions is to be met'. Table 8 shows the strength of various classes as defined in British Standards 187 and 3921.

BS 3921 defines bricks of special quality as 'Bricks that are durable even when used in situations of extreme exposure where the structures may become saturated and be frozen, eg, sewerage plants, retaining walls and pavings'.

In *Brick floors and brick paving*,[49] Thomas and Bovis qualify the need to use bricks of special quality by writing 'It is not essential that pavoir bricks be of special quality. Other bricks which fail to meet some of the requirements of

TABLE 8 Brick strength as defined in British Standards 187 and 3921

Type	Minimum average compressive strength N/mm²
Clay engineering class A	69.0
Clay engineering class B	48.5
Clay loadbearing from class 1	7.0
to class 15	103.5
Calcium silicate* from class 4	27.5
to class 7	48.5

Calcium silicate classes 1 to 3 not shown as not recommended for external paving
Source: Architects' Journal, *17 September 1975*

"special quality" (as defined in BS 3921) may give excellent service and if designers have any doubts about their suitability they should consult the manufacturer requesting advice and/or evidence that the bricks have given satisfactory service under conditions similar to those proposed'.

Most bricks which are not manufactured especially as pavers must be laid on edge if their 'normal' face is to be exposed. They can however, especially if they are of the solid wire-cut type, be used in the 'on-flat' position.

Clay pavers These units are specially manufactured to be laid flat as paving. They are produced in a variety of rectangular sizes and in thicknesses which extend from 20mm upwards (see Table 9 below). The colour range is more limited than for bricks but it is nevertheless fairly extensive and includes grey, charcoal, buff, dark brown, red and blue.

Pavers which are made especially to withstand heavy wear are obtainable with a variety of mechanical surface textures, as illustrated in **9.53**.

Calcium silicate bricks Bricks to BS 187, Class 4 or higher, should be selected when calcium silicate bricks are used for road paving.

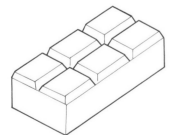
panelled

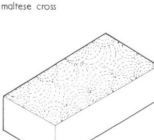
maltese cross

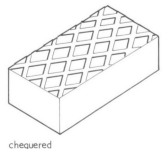
chequered

dimpled

9.53

BRICK SIZE The metric brick 'format' (ie, brick + joint) quoted in BS 3921 is 225 × 112.5 × 75mm. Subtracting a 10mm joint, the 'work size' (ie, the actual size of the brick) is 215 × 102.5 × 65mm.

'Modular size' bricks are also made by some manufacturers. The format sizes of these include 300 × 100 × 100mm, 300 × 100 × 75mm, 200 × 100 × 100mm and 200 × 100 × 75mm.

Both standard bricks and modular bricks are used for paving as are pavers, already described. Pavers are thinner than ordinary bricks but the other dimensions may be different as well. Table 9 below lists various brick and paver dimensions.

TABLE 9 Commonly available sizes of bricks and pavers and number of units per square metre of paving allowing for 10mm joints (no allowance for cutting or waste)

Type	Unit work size (mm)	Exposed face format (mm)		No per m²	m² per 1000
BS bricks	215 × 102.5 × 65	On edge	225 × 75	60	16.66
		on flat	225 × 112.5	40	25.00
Metric modular bricks	290 × 90 × 90		300 × 100	33	30.30
	290 × 90 × 65	On edge	300 × 75	44	32.73
		on flat	300 × 100	33	30.30
	190 × 90 × 90		200 × 100	50	20.00
	190 × 90 × 65	On edge	200 × 75	66	15.15
		on flat	200 × 100	50	20.00
Pavers	215 × 102.5 × 20 33, 40, 50, 55, 60		225 × 112.5	40	25.00
	215 × 65 × 33		225 × 75	60	16.66
	215 × 140 × 33		225 × 150	30	33.33
	190 × 90 × 50		200 × 100	50	20.00
	250 × 125 × 50		260 × 135	28	35.71
	300 × 150 × 50		310 × 160	20	50.00

Source: Architects' Journal, *17 September 1975*

PATTERNS OF LAYING
The bond pattern is of great importance in determining the appearance of the brick surface. A large choice is available. The herringbone pattern is the best for resisting the forces imposed by vehicles but when the loading is light others can be chosen.

It should be borne in mind that when the joints run at right angles to the direction of traffic slipperiness is reduced.

Bricks may be used on edge instead of flat although this is a more expensive usage. The scale of the surface, for any particular bond, will also be significantly affected by the size of brick used. Sketches **9.54** and **9.55** illustrate a number of possibilities.

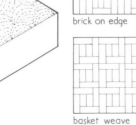
brick on edge · brick on flat · basket weave · brick on flat

9.54 *Basket weave variations*

9.55 *Other variations*

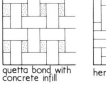

stretcher bond

diagonal stretcher bond

quetta bond with concrete infill

herring bone

9.57 *Typical section of road pavement*

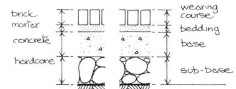

brick
mortar
concrete
hardcore

wearing course
bedding
base
sub-base

9.56

The brick surface of this service vehicle access road **9.56** has the appearance of being integral with the containing brick side walls. The stretcher bond is laid transversely providing maximum traction and the expanse of the road is divided by the drainage channel along the centre line, its linear quality being accentuated by the direction of the courses.

THE ROAD PAVEMENT The design of sub-base and roadbase should be in accordance with normal engineering design procedure. The sub-base will be of consolidated hardcore or other suitable material. Roads which take other than very light vehicular traffic should have a concrete roadbase, reinforced or unreinforced according to the requirements of the design. A typical section is shown in **9.57**. For very lightly trafficked roads and for pedestrian ways it is necessary to decide whether to use a concrete or a consolidated hard-core roadbase. In making this decision the possibility of service vehicles, for example fire engines, using a pedestrian way must be taken into account.

In view of the small size of the units the possibility of differential settlement should also be considered in particular from the points of view of safety, appearance, drainage and cost of repair. In the UK it appears to be usual to adopt a concrete base in urban situations even when the route is solely for pedestrians.

On a concrete base the bedding can be of sand or mortar, although for vehicular traffic a sand bedding does not provide resistance to the forces causing lateral movement and so mortar is preferable unless the loading is very light, and in fact mortar bedding is normally used with a concrete base.

The thickness of mortar bedding is usually 25 mm. Sand bedding on concrete should be at least 25 mm thick and on hardcore 50 mm should be used.

A suitable bedding mix is 1 part Portland cement, $\frac{1}{4}$ part lime and 3 parts sand by volume. A 1 : 4 lime/sand bedding is sometimes used over hardcore.

The units can be butt jointed, or the joints can be filled with sand or mortar. The choice will depend on type of traffic, required appearance, and whether or not a porous surface is necessary. For vehicular traffic mortar joints should be used. These are usually 10mm thick but for particularly irregular bricks they may be larger. Consideration can be given to the use of colouring additives to the mortar.

Butt joints are rarely used because of irregularities in the units. Mortar jointing can be carried out as the bricks are bedded or the joints can be left open and filled later either with grout or a dry mix which is then wetted by spraying. It is important that the jointing process be carried out with great care by skilled workmen using an established procedure otherwise the paving will be disfigured by the mortar.

For grouted joints or joints filled as the bricks are laid a 1 : $\frac{1}{4}$: 3 cement : lime : sand mix by volume should be used. For a dry mix 1 : 3 cement : sand is usual.

Surface drainage   Falls in the range 1 : 40 to 1 : 60 should be used.

Movement joints   The joints in the paving should coincide with those in the concrete roadbase. Movement joints should also be used where the road surface abuts building structures, retaining walls and other impediments to expansion and contraction.

Edge restraint   Edge restraint is required when the bricks are bedded in sand. With bedding in mortar it is not necessary. Some support, however, will be needed for edge bricks particularly if wheel loads can come to the edge of the surface.

Special units   Perforated bricks are now used more often in situations where a self-draining surface is required. They can be a useful paving material for hard standings and parking areas.

The possibility of prejointed brick paver units being adopted is under investigation and this could become an economical alternative to the normal brick pavers by providing a method of paving which is both quick and minimizes the dangers of staining during construction.

Safety   Bricks should be chosen for anti-slip characteristics appropriate to the proposed use. Many bricks suitable for paving have excellent anti-slip properties and the joints contribute to this characteristic. Some brick pavers are manufactured with a dimpled surface. The bricks most prone to be slippery are those which are most dense, usually ones having high crushing strengths.

Colour and the road pavement

The horizontal, ground level elements in a townscape view are secondary to the vertical elements in visual importance and therefore should not be obtrusive. As a consequence, in most cases of townscape alignment, colour should be used in a restrained manner with subtle differences and harmonies in tone, shade and colour being preferable to marked colour contrasts. Strong colour differences easily appear strident and can soon be experienced as an imposition if they are constantly in view. Harmonious colours are more restful than contrasting ones and so the latter are appropriate for residential areas whereas a stronger usage might be called for in, say, a shopping mall or precinct.

Flowing alignments require a different and occasionally more bold use of colour in scale with the speed and size of the road. By employing a different coloured surface for the hard shoulder from that of the carriageways the visual impact of the motorway width can be reduced. Bright colours on deceleration lanes or motorway run-off ramps can provide early warning as well as visual interest. The transverse yellow lines used to induce reduction in speed on motorways have a decorative quality as well as fulfilling an engineering function, and in fact all road markings provide an opportunity to enhance the environment by adding colour and variety to the surface. There is scope for greater use of areas of colour rather than the linear uses which are dominant at the present time when it is required to convey some instruction to the road user. Again discrimination is necessary; the road must look satisfactory to those who can see it from outside the carriageways as well as the road users themselves. It is difficult to imagine a justifiable use of two separate colours for each lane in a two lane carriageway instead of a white line down the middle, but to have a different coloured surface at a roundabout and perhaps a short distance down the approach roads might sometimes be useful in warning the driver of a potential hazard.

Visibility is improved with lighter coloured surfaces. This is especially true for night travel. So there are a number of engineering reasons as well as aesthetic ones which make

colour control advantageous.

The use of coloured aggregates in the construction of both asphaltic and concrete roads is a possible means of obtaining a coloured surface finish. A useful document in the selection of an aggregate is *Sources of White and Coloured Aggregates in Great Britain*, prepared by the Road Research Laboratory and published by Her Majesty's Stationery Office.[50]

The two main classifications of artificial stone (electric and blast furnace slag and crushed brick) and natural stone (quarried rocks and gravel) are described in BS 812, entitled Methods for sampling and testing of mineral aggregates, sands and fillers. Of course, any stone selected must be adequately tested for suitability.

The most strongly coloured stone is the result of geological decomposition. Since such usable deposits normally occur surrounded by rock which is not suitable for roadstone, those with the strongest colours tend to be expensive. However there are enough less highly coloured stones available to provide a reasonable choice and, in any case, these are the ones likely to produce the most pleasing results.

Colouring additives are available for both concrete and asphaltic materials. These should be used with discrimination to ensure a satisfactory weathered appearance.

A range of coloured interlocking concrete paving blocks is available.

Colours include natural grey, buff, red, dark grey, yellow, ochre, browns, purple and blue.

Excellent results can be achieved by the use of bricks or brick pavers on the road surface. Colours available include reds, orange, yellow, ochre, browns, purple and blue.

ASPHALTIC PAVEMENTS

These are constructed with a binder of natural asphalt, a distillate of petroleum which has been condensed to form bitumen, or a distillate of coal condensed into tar. All these materials are black or very dark brown in colour.

Binders with a plastic base and others utilizing synthetic rubber are coming onto the market and may eventually greatly alter the options with respect to colour. However in the meantime there are a number of ways in which the impact of the traditional, predominantly black, binders of asphaltic pavements can be modified or even negated. On a well trafficked road the surface of the aggregate used to construct the asphaltic wearing course will eventually be exposed by the wearing of the bituminous binder. Removal of the surface binder is helped by the process of weathering since oxidation of the binder occurs. A judicious choice of aggregate can therefore control the subsequent surface colour. Alternatively a selected, coated aggregate can be rolled into the surface. This will be exposed in the same manner but has the advantage of minimizing the quantity of what could be a more expensive aggregate than that used for the main construction of the wearing course. Uncoated coloured aggregate can be rolled into the asphaltic surface. Natural gravel or broken stone aggregate can be used for this purpose and a large range of effects can be produced depending on the colour, shape and size of the aggregate and the rate of spread. This technique is usually reserved for less heavily trafficked roads where the speeds are relatively low.

The application of surface dressing is a standard practice which has been described above under the heading The Asphaltic Pavement. A normal method of application is to spray the surface with a tar or bitumen binder and then spread and roll in the aggregate. Clearly surface dressing is another way in which the surface colour and texture can be altered. Dyes can be added to the wearing course mix to produce a range of colours. The choice is limited as a consequence of the black binder but nevertheless includes a useful range of reds and browns. Green can also be obtained but is less common and is also much more difficult to use as a successful colour component in a design. The dye most often used is a heat-stable iron oxide based pigment which is specially prepared for this purpose. It is normally added at the rate of 1–3% of the total material.

Slurry seals and bituminous or resin-based paints are available for adding colour to bituminous surfaces. It is only practical to use them in lightly trafficked areas. These liquid sealers are applied in two or three coats and quite a wide range of colours is available.

CONCRETE PAVEMENTS

The coarse aggregate, the fine aggregate and the cement all play a part in determining the colour of a concrete surface. If the coarse aggregate is not to be exposed by some finishing technique then the fine aggregate and the cement will be the most important ingredients in this respect, but if the aggregate is exposed then it must be selected for colour and texture as well as its engineering qualities. The best results will be obtained when the three constituents are chosen to complement each other in attaining the desired finish.

White concrete can be produced by using white cement in conjunction with a light coloured aggregate. The range of satisfactory aggregates available is limited. Those which are suitable include the lighter coloured limestones and granites, calcite spar, calcined flint, and one or two deposits of natural fine silica sand. Also available are imported white aggregates. These are generally sold under a trade name but normally consist of gabbro or quartzite. Further details can be obtained from *White Concrete: with some notes on black concrete* by J. Gilchrist Wilson.[51] Appendix 1 of that publication lists some light coloured aggregates suitable for the production of white concrete. The name and type of aggregate are given together with the name and address of the supplier. Equivalent details are given in Appendix 2 for some dark or very dark grey aggregates suitable for use in the manufacture of black concrete.

Designers should bear in mind that a primary attribute of a good road design is that it blends with its surroundings. It should be an integrating rather than a divisive element. Light coloured surfaces are only appropriate when their use is compatible with this end.

In the construction of coloured concrete road surfaces particular care must be taken to ensure consistent results. Whenever possible test panels should be made and a strict procedure for the preparation of the mix and the construction of the slab established. It is necessary to ensure that sufficient fine and coarse aggregate of the required colour is available to complete the work and regular inspection and comparison with a retained sample should be carried out. All the coarse aggregate should be obtained

from a single source, as should the fine aggregate. Ideally the part of the pit or quarry from which the aggregate is obtained will not be changed during the contract. Consideration may also be given to stockpiling sufficient aggregate to complete the work. If this is done it is necessary to protect the stockpile from variations in water content and from contamination.

If the shape and grading are not also controlled there is a danger of differences in surface colour as a result of alterations in the mix proportions.

The use of coloured cements or pigments is also an important method of obtaining a coloured finish. A range of colours and shades are available as well as white and black. By mixing white cement and black pigments with ordinary Portland cement shades of grey can be produced.

It is advisable to give cement manufacturers adequate notification of quantities to be supplied. If it is necessary to use more than one manufacturer for the provision of white cement this will not necessarily be a disadvantage since the whiteness is not likely to vary between one manufacturer and another.

Black cement can be produced by adding black pigment to ordinary Portland cement. Although pigment can be obtained for addition on site it is very difficult to achieve good results using this method (this applies for any colour of pigment). If pigment is used it should be added during the cement manufacturing process. In the construction of a slab with black concrete it is necessary to expose the aggregate if a satisfactory finish is to be produced; the reason being that a smooth, black surface is likely to appear blotchy and it shows up lime bloom. Also, since the pigment used to blacken cement is very fine it tends to come away on the surface.

The colours available in cements range from pink and buff through reds, browns, light green and blue. These cements are manufactured by the addition of powdered pigments or paste pigments to ordinary Portland cement, white Portland, or other Portland cements.

Coloured cements should be used with discretion. When the aggregate is not exposed the best results are likely to be obtained with pastel shades; when the aggregate is exposed the colour of cement should be chosen in conjunction with that of the aggregate.

Pigments are specified in BS 1014, Pigments for Portland cement and Portland cement products. Up to 10% pigment in relation to cement is added depending on the colour and shade required.

10
LINES, AREAS AND TEXTURES

Lines

In railways the use of line as a design element could not be more obvious. The materials and form of the railway line are completely determined by the characteristics of the vehicles. By properly fulfilling their function the lines attain their own intrinsic beauty, **10.1**.

The use of line in road design is a more complicated matter. The running surface is an area but one which is elongated to such an extent that it exhibits many of the characteristics of a line. Road line at this level has been considered in Part I, The Townscape Alignment and Part II, The Flowing Alignment. Here we will look at the local use of line when superimposed on the road surface.

The ways in which line can be employed locally are many. It can fulfil a purely functional purpose (the dotted line separating carriageways), it can attain a purely decorative end (the patterned surface in a shopping precinct) or it can fulfil a specific function at the same time as its visually pleasing qualities are being exploited. This latter usage is often the most satisfactory.

THE DELINEATION OF EDGES

Lines can be used to delineate the edge of an area and thereby clarify the form. In terms of roads this entails outlining the boundary between two surfaces having different usages and often consisting of different materials. To achieve a pleasing result the shape of the line should be in harmony with the character of the areas it separates. This has been achieved in the informal example of **10.2**, which depicts the edge of one of the parking areas off an access road at the Brow, Runcorn, Cheshire. In this example the line is relaxed and pleasing and the materials go well together.

In the example of **10.3**, where a simple geometric configuration has been used, the result is much less pleasing. Line has not been used as an integral part of the design but solely as an engineering concept relating to the turning circles of vehicles.

10.1

10.2

10.3

257

10.4

10.5

LINE AND DIRECTION

Two ways of using line to emphasize direction are illustrated in **10.4** and **10.5**. In **10.4** the linear characteristic of the road edge has been exploited, whereas in **10.5** the unit paving has been orientated along the direction of flow of the road surface, so that the surface itself becomes an assembly of parallel lines. This procedure is one of the factors which has enabled the designer to dispense with an upstanding kerb.

Areas of special usage need to be defined. Ideally a designated usage should be apparent as a result of an integrated design rather than by means of signs.

The use of lines is one way of conveying the necessary messages. In **10.6** they have been employed to circumscribe the parking bays, and in **10.7** they indicate a large scale pedestrian crossing.

10.6

10.7

258

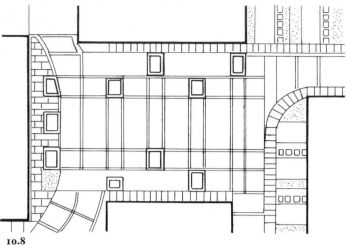

10.8

LINES AND AN INTEGRATED ENVIRONMENT

Lines can be used to assist in the integration of the road surface with the other elements of the surrounding environment.

In **10.8** and **10.9** the line of brick pavers inlaid into the asphaltic pavement surface are set out to the same module as the dwellings, thus linking the surface to the building elevations. At the same time the planting and road pavement are unified since the planting boxes are constructed to fit the grid of pavers. (See also Millers Court, Hammersmith, illustrated on p. 19.)

LINES AND PERSPECTIVE

The view of a surface on a 1:200 plan can be very different from that seen by a person walking across it. There are a few simple characteristics of lines as seen in perspective which it is useful for a designer to bear in mind when he turns his attention to the road surface.

Lines which are parallel in plan, when viewed along their length appear as though, if extended, they would converge at a single point on the horizon. The nearer the observer is to the lines, the longer they will appear. The nearer the observer is to the lines the wider apart they will appear to be in relation to the observer. These points are demonstrated in

10.10. Horizontal lines which are parallel in plan, when viewed at right angles to their length will continue to appear parallel. If the lines are equally spaced in plan those furthest from the observer will appear closest together in perspective. The nearer the observer is to the lines the wider they will appear and the further apart they will appear.

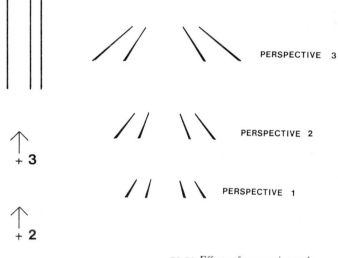

10.10 *Effects of perspective and distance from the observer on horizontal parallel lines when viewed along their length*

10.9

259

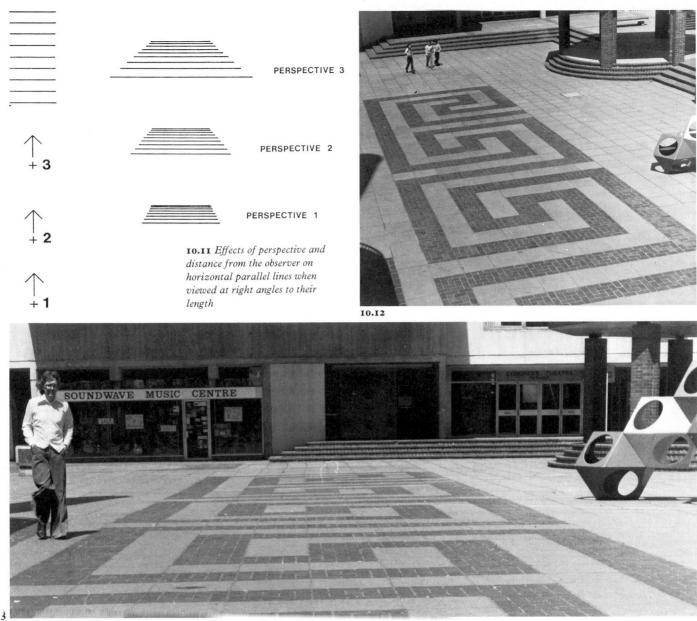

PERSPECTIVE 3

PERSPECTIVE 2

PERSPECTIVE 1

+ 3

+ 2

+ 1

10.11 *Effects of perspective and distance from the observer on horizontal parallel lines when viewed at right angles to their length*

10.12

10.13

The effects indicated in **10.11** are illustrated by the example of photograph **10.13**. The height of the observer above the surface also has a great influence on the way he sees lines and patterns, as can be seen by comparing **10.12** and **10.13**. This factor should be considered when it is possible to view the surface from elevated walkways, balconies or tall buildings in the vicinity.

A circular curve when seen in perspective becomes an ellipse, **10.14**. It is obvious, but important, that this is true from whatever angle the circle is viewed.

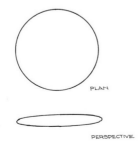

PLAN

PERSPECTIVE

10.14

PLAN

PLAN

PLAN

3

2

1

PERSPECTIVE

3

2

1

PERSPECTIVE

+ 3

+ 2

+ 1

10.15

A circular curve on a horizontal surface when viewed at right angles takes on more and more the appearance of a straight line the further the distance of the observer (and the lower his viewpoint).

A circular curve on a horizontal surface when viewed along its length, appears to turn a more and more sharp angle the further the distance of the observer (and the lower his view-point).

10.16

LINES AND THE THIRD DIMENSION

The linear aspect of roads can be emphasized by designing associated vertical elements in harmony with the alignment. Two examples of this are shown in **10.16** and **10.17**. In **10.16** a delicate welded steel railing and in **10.17** a well proportioned brick wall are an integral part of the designs.

10.17

10.18

LINES AND PATTERN

A traditional use of line on the floorscape is the decorative one of creating a pleasing pattern. This is often associated with adjusting the scale of the surface to harmonize more with the human form.

Since scale and speed are closely associated, by reducing the scale of an area in this way it can be implied that the appropriate speed to traverse the surface is a walking pace rather than the higher speeds attainable in a car.

An example of the use of lines to create pattern and reduce scale is shown in **10.18**. This example is from a precinct allocated to pedestrian use only, but such a surface could be equally appropriate in some areas with vehicle access.

10.19

Areas

Pattern can be created on the surface by sub-division into distinct areas as well as by the use of line. As with line, areas can alter the scale of the surface as well as introduce variety and interest. In fact sub-division into areas can be employed to indicate or control edges, direction, motion and usage. It can in other words fulfil the same functions as line as well as being a source of applied decoration in its own right.

One of the most common applications of sub-division into areas is that of designating surfaces for parking; photograph **10.19** illustrates the use of interlocking precast concrete paving blocks to indicate the parking area adjacent to an asphaltic running surface.

262

Different colours or shades of concrete paving blocks can be used to indicate the exact parking positions for cars on a surface composed entirely of blocks, **10.20**.
Another example of the perspective view of a geometric pattern, **10.21**, is shown in **10.22**, in which the pattern is composed of areas, rather than lines. Again the example is taken from a pedestrian precinct but it serves to underline the point that the bird's eye (drawing board) view is quite different from the view at ground level.

An intriguing aspect of the way in which patterns are seen on the ground is the contrasting appearance that can be presented from different viewpoints. When seen from direction A the forms of **10.21** are truncated and attention is drawn to the diagonal element in the pattern, whereas when seen from direction B they appear to be elongated and the horizontal, linear aspect is dominant. These aspects of pattern should engage the attention of the designer as he develops an arrangement for a particular location.

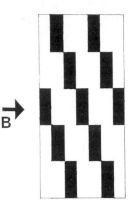

10.21

10.20

10.22

10.23

263

Texture

The variety of texture available to the designer, whether using a flexible or rigid pavement or unit paving, has been described earlier in this chapter.

Texture is a third tool, in addition to line and area, which in the hands of an imaginative designer will facilitate the enrichment of the road surface in a meaningful way. It can achieve similar objectives to those listed for line and area. Two particular aspects of their use are mentioned below.

TEXTURE AND MOVEMENT CONTROL

The texture of some surfaces are pleasant and blend easily with their surroundings whereas others convey the message 'keep off'. The rougher the texture the more it inhibits movement. This characteristic is enlisted to control the movement of vehicles and pedestrians across the surface.

10.25

Areas which are to be prohibited to both vehicles and pedestrians will be given the roughest texture. Cobbles have often been used for this purpose, **10.24**. Pedestrians can be dissuaded from crossing a road at an unsuitable place if they are faced with a hazard of this type on the other side. A surface can have a degree of roughness which is sufficient to discourage pedestrians but at the same time enables vehicles to cross it at very low speeds without causing undue discomfort to the occupants. This expedient can be employed, as in **10.25**, at the entrance to garages, with the obvious advantage that vehicle-pedestrian conflict is unlikely in an area where it could be particularly hazardous.

Another option which could be open when dealing with short access roads is that of making the entire running surface sufficiently rough to necessitate slow speeds. This option has been adopted in the example of photograph **10.26**. In this case the road surface has been constructed with flat stones embedded in hoggin.

10.26

10.24

10.27

10.28

A more common expedient encouraging low speeds is that of constructing a localized area of rough surface across the road width, **10.27**.

On exceptionally steep slopes roughened surfaces can greatly assist the manoeuvrability of vehicles. Photograph **10.28** illustrates the in-situ concrete tracks which were necessary on a steeply sloping site when constructing the approaches to each individual garage. As can be seen, the concrete was deeply indented by pressing in the edge of a board before the final set.

Movement control can be exercised also by the adoption of materials which are less rather than more rough than the normal running surface, **10.29**. The use of materials which appear more 'fragile' or which are normally associated with the inside or immediate surroundings of buildings rather than the open highway will induce caution on the part of the motorist.

10.29

10.30

RELATING TEXTURES

Some materials look well together and others less so. The object should always be to use in association those which will enhance each other's appearance, at the same time achieving an overall harmony. There are no simple rules whereby successful results can be guaranteed; only observation and experience will enable the designer to develop his skill in this respect.

The location in which a material is used, including the general character of the surroundings, the functions which the surface must fulfil, the proportional area of one material in relation to another are all factors which will affect the success of a particular association. A few examples of juxtaposed materials are presented below.

The laying of an asphalt finish on wide footpaths bordering a wide road with a bituminous macadam surface tends to produce a rather overpowering black asphaltic area, **10.30**. In this example the situation is somewhat alleviated by the light coloured chippings rolled into the surface of the pavement, but the drop-down kerbs give a look of restlessness to the already rather forbidding hardness of the view.

Precast concrete slabs and a kerb of dark coloured setts, **10.31**, laid almost flush with the dark macadam road surface present a harmonious whole, with the vehicle and pedestrian areas clearly defined. A precast concrete sleeping policeman straddles the running surface.

10.31

266

10.32

10.33

In **10.32**, precast concrete slabs, a stone upstanding kerb and interlocking concrete paving blocks form a pleasing composition. The blocks are in two tones laid in alternate rows. Small units like these do not show up the oil stains as much as in-situ concrete with a smooth finish or a precast surface of large flat units.

In-situ concrete with exposed aggregate, smooth-surfaced brick pavers, pavers with diamond pattern, precast concrete channel units and fine textured macadam have been skillfully combined to produce the result depicted in **10.33**.

11
TRIM

The functions of trim are to protect surface edges and retain the layers of a pavement, to mark or protect boundaries, to mark junctions between surface materials, sometimes to form construction joints, to collect surface water and to control traffic. In addition it has a decorative function to perform and in this respect it should always be used in a positive manner and not simply considered as necessary in order to fulfil a particular engineering requirement.

It is important in the choice of trim materials and details that stock solutions are not automatically adopted and that every effort is made to use and, when necessary, develop details which are in character with the immediate surroundings and with the environment of the neighbourhood. By its form, texture and colour the trim can greatly enhance the amenity of outdoor spaces. Relative levels are important. The trim may protrude a small amount above the general level, like deterrent surfaces or sleeping policemen, or a large amount like bollards; again it may be flush like rows of pavers or setts marking parking bays or a pedestrian area, or these may be depressed to form a gutter. Upstand kerbs may be made from granite, whinstone, sandstone, reconstituted stone, precast concrete or brick. Flush or depressed trim may utilize these materials as well as cobbles embedded in concrete, setts, in-situ concrete, asphaltic and loose materials including gravel, larger stones and loose cobbles. There are numerous sources of good details (see the bibliography at the end of the book).

In selecting materials for trim the initial cost should be considered in relation to durability and maintenance costs. If this is done the use of more expensive materials is often justified by these factors alone and there is the added amenity benefit of high quality. In this section a few examples are presented of trim which has been carefully designed to be part of the total streetscape.

Edging, kerbs and gutters

The importance of the road edge detail cannot be over emphasized. It plays a major role in determining whether the road surface is to become an integral part of the scene or a gash of concrete or asphalt cutting off one side of a space from the other. This detail is as critical in the country scene as it is in the town or suburb. Inter-urban motorways have flush kerbs at the edge of the carriageway since upstand kerbs are dangerous at high speeds and they would prevent a vehicle from pulling off the carriageway in an emergency. Well designed country roads also often have flush kerbs or they are omitted altogether, it being considered adequate to extend the roadbase 300mm beyond the edge of the running surface. In the suburban and other townscape contexts the 100mm upstand, precast concrete kerb appears to be almost universally adopted at the present time. Where pedestrians must walk adjacent to vehicles travelling at speeds which could constitute a danger for those on foot some separating device is required. An upstand kerb is one way of attaining this separation, but there are alternatives to the 100mm high light grey concrete kerb. Now that the needs of pedestrians are being given more recognition in the townscape setting, there is broader scope for varied solutions. Kerbs are constructed to mark the edge of the carriageway, to prevent lateral spread of the

II.I

pavement, to form an abutment for the control of surface water drainage, and to protect pedestrians and property. The design of a particular kerb should be considered in relation to these functions. It will often be found that a kerb is not needed to fulfil one or more of the functions listed. If a concrete road is to be built a kerb is not required to contain its sides. If a dual use area is being designed then a kerb is not required to delineate the pedestrian space. (An excellent dissertaion on kerbs is presented by Elizabeth Beazley in *The Design and Detail of the Space between Buildings*;[52] this book is recommended for all aspects of the subject covered by its title.)

Grey precast concrete kerbs are found in most developments throughout the country, but the use of a darker coloured kerb is less divisive, **II.I**. Trim should always be chosen to enhance the appearance of adjacent surface materials. The lighter colour is more visible at night but in residential areas with slow moving traffic this may not be an overriding consideration.

A single row of setts slightly raised above the road level clearly delineates this footway, **II.2**. The use of a rough textured asphaltic surface with fine gravel scattered on top imparts an informal air, in keeping with the local village setting.

An informal edging to a court pavement is shown in **II.3**. Here the setts, built up at an angle to contain the planting beds, have a similar texture to the interlocking blocks of the road surface.

A double kerb is convenient for the pedestrian when the footpath is sufficiently elevated to make it feasible, **II.4**.

II.2

II.3

II.4

Sleeping policemen

It is not yet lawful to incorporate sleeping policemen in a public highway. However the matter is at present being investigated by the Road Research Laboratory and it may not be very long before the legal position is altered. In the meantime they are used to good effect on private roads and in other special circumstances. They are usually constructed in a material different from that of the general road surface in order to render them more visible to the motorist.

11.9

A narrow precast concrete upstand has been used in this cul-de-sac, slowing vehicles before they are parked in the small, well concealed groups of hardstandings on the right-hand side of the road 11.9. Instead of an upstand a small drop of say 50 mm can be incorporated across the width of the road to act as an inducement to reduce speed. A suitable location for such an obstacle might be across the entrance to a mews court.

In 11.10 a wide band of brick has been placed across the road at the level of the footpath continuing the line of the pedestrian precinct within the housing area.

Sometimes all that is required to induce increased caution on the part of the driver is a rumble strip of material with a coarser texture than that of the general running surface. 11.11.

11.10

11.11

273

11.12

11.13

Bollards

In relation to roads, bollards are used to prevent vehicles encroaching on pedestrian areas, to protect property and to indicate divisions in territory. They can take the place of a kerb when footway and road are on the same level and they can be used to prevent vehicles parked at right angles to the kerb from overlapping the footpath. Traditionally bollards have been used to protect the corners of buildings, gateposts and other parts of properties vulnerable to damage from vehicles.

Removable and hinged bollards are useful for excluding unauthorized vehicles from special areas. These are often employed at the entrance to access ways which are primarily for the use of pedestrians but which can also be for the benefit of service vehicles. By keeping the bollards locked in position private cars are excluded but the service vehicle driver with a key can unlock and remove or lay flat the bollards and gain access. Hinged bollards are also used to preserve hard standings for the exclusive use of the person with the key to the locking device. Illuminated bollards sometimes afford a pleasant alternative to overhead lighting. They are also useful for the direction of pedestrian and vehicle traffic.

Reinforced concrete, reconstituted stone, timber and steel bollards are all readily available. Traditionally, natural stone (often granite or sandstone) was used. These are now declining in use owing to their relatively high cost but they will be supplied to order by stone firms. Cast iron, another traditional and very attractive material for bollards is also going out of use and foundries are ceasing to manufacture them from the traditional patterns. However there is evidence of a revival of interest in the material and bollards will be cast to order.

Cast iron out of doors is normally bonderised and then painted but hot-dip galvanising and metal spraying might be considered. Mild steel combines lightness with strength. Care should be taken to ensure that the proportions are such as to make steel bollards look strong and in scale with their surroundings and function. Fixed steel bollards should have a flat plate welded to the bottom complete with bolt holes. The removable type are usually slotted into a hole formed by a sleeve set in the pavement. These sleeves are available as standard items. Steel bollards must be painted and preferably galvanised for additional protection.

Timber bollards can be of hardwood or softwood. For the most pleasing visual effect they are likely to require more substantial cross-sectional dimensions than would be demanded on considerations of strength alone. Suitable hardwoods include oak (excluding sapwood), afrormosia, ekki, greenheart, jarrah, makore and opepe, none of which require preservative treatment. Softwoods should be treated with a wood preservative and the part which will be buried should be treated with creosote. They should also be free of large knots.

If a change in colour is required it is best achieved with wood stains, in which case the wood will need less maintenance and will last longer than if it is painted or varnished. Stains have the additional advantage that the grain remains visible.

A good choice of simple pleasing shapes in precast concrete bollards (some with exposed aggregate finish) is commercially available. The one on the left of photograph 11.12 is interposed between the footway and the main area of the dual use court. That on the right indicates and protects the corner of the brick wall. The option of forcibly restricting vehicle access to the area has been reserved by the installation of a sleeve for a removable bollard mid-way between the two permanent ones.

Well proportioned timber bollards create an unequivocal barrier to vehicles, 11.13. Steel and cast iron bollards look well in the highly urban setting.

THE DYNAMICS OF ALIGNMENT

Contents

A road alignment must be safe and comfortable to drive along at the design speed. When traversing curves an adequate factor of safety against skidding is required, sight lines both in the horizontal and vertical planes must be sufficiently long for the design speed and the performance of the vehicles, and at the same time the driver should not experience any sense of unease from the effect of centrifugal force. The effects of this force are two-fold. It tends to cause the vehicle to skid and it tends to pull the driver sideways from a comfortable driving position.

A motorist will normally travel on a curve at such a speed that he does not experience any significant degree of discomfort due to the centrifugal effect. In the flowing alignment the radii of bends should be sufficiently large that they can be negotiated with ease at the design speed. Also the speed which does cause discomfort should be well below that which causes skidding. The driver will then be warned in sufficient time that he is driving too fast for the road curvature. At the same time the designer should consider using the choice of radius as a means of discouraging drivers from exceeding the design speed. In other words the radii may be sufficiently small that the driver begins to feel discomfort if he exceeds that speed.

All the factors mentioned—speed, radius of curvature, centrifugal force, resistance to skidding, the sight lines—relate to the dynamics of vehicle movement and largely determine the scale and form of the flowing alignment. It is necessary for the intelligent preparation of satisfactory alignments that the designer understand the relationships between these factors. This is the topic of the following paragraphs.

1 The horizontal plane

CURVED ALIGNMENTS—THE RELATION BETWEEN CENTRIFUGAL FORCE (F) MASS (M) RADIUS (R) AND VELOCITY (V)

When a vehicle follows a curved path at constant velocity it is actually imparting an acceleration to its mass, since acceleration is the rate of change of velocity, and in following a curved path the direction of the velocity of the vehicle is changing.

Assume a vehicle is travelling round a curve of radius R with constant velocity v and that it moves from point A to point B, **A.1**.

if θ = Angle denoting change in direction of travel between point A and point B

    V = Vector change in velocity

    S = Distance AB

then $V = v\theta$ and $\theta = \dfrac{S}{R}$

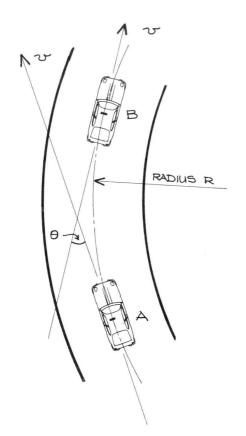

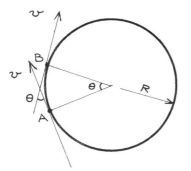

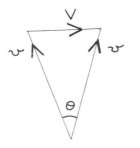

A.1

$$\therefore \ V \ = \ \frac{vS}{R}$$

$$\frac{dV}{dt} \ = \ \frac{d}{dt}\left(\frac{vS}{R}\right) = \frac{ds}{dt} \cdot \frac{v}{R}$$

but $\quad \dfrac{ds}{dt} = v$

so $\quad \dfrac{dV}{dt} \ = \ \dfrac{v^2}{R} \ = $ acceleration

Since force = mass × acceleration therefore the centrifugal force acting on a vehicle of mass M is given by (Equation 1):

$$F \ = \ \frac{M}{R} \times v^2$$

In Equation 1, for a particular vehicle,

(a) if v is constant $\qquad F = \dfrac{\text{constant}}{R}$

(b) if R is constant $\qquad F = \text{constant} \times v^2$

(c) if F is constant $\qquad R = \text{constant} \times v^2$

We will now look more closely at these relationships between the constituent elements of the equation. (A low radius of curvature has been used in the examples for (b) and (c) to explain the points being made. Such radii would not, of course, be used in flowing alignments for major roads. It should also be noted that the examples and graphs in this appendix are presented with a view to elucidating the principles involved and not for direct use in design problems.)

(a) *Relation between R and F when v is constant*
The relationship, centrifugal force,

$$F = \frac{\text{constant}}{\text{radius of curvature, R}}$$

for the case when the velocity is 48 km/hr (30 mph) and the vehicle weight, W, is 15 kN ($1\frac{1}{2}$ tons) is shown in **A.2**. This curve is a hyperbola.

It can be seen from the graph that when the radius is small, a small change in radius effects a large change in the centrifugal force. When the radius is large then a large change in radius brings about only a small change in the force.

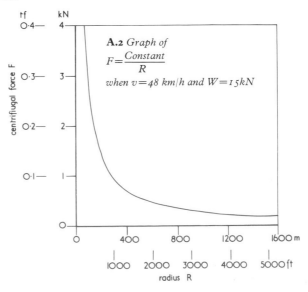

A.2 *Graph of* $F = \dfrac{Constant}{R}$ *when* $v = 48 \ km/h$ *and* $W = 15kN$

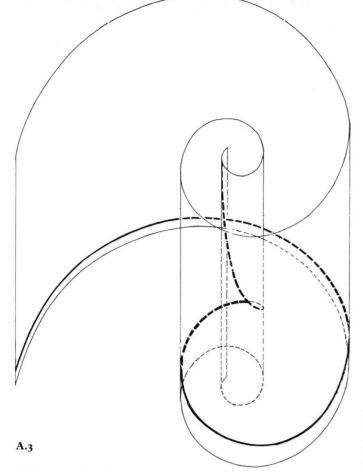

A.3

This graph has been plotted isometrically on a spiral base in **A.3**. The radius of the base line changes at a constant rate so the relationship between centrifugal force and radius can be easily seen.

It is apparent that the use of a horizontal transition curve with a constantly changing radius can bring about a marked increase or decrease in the centrifugal force. The abruptness of the change in force at any particular speed will depend on the rate of change of radii of the transition curve and the range of radii covered by the curve.

(b) *Relation between F and V when R is constant*
The relationship, centrifugal force, F = constant × (velocity)² for the case when the radius is 150 m (492 ft) and the vehicle weight is 15 kN (1½ tons) is shown in **A.4**. This curve is a parabola.

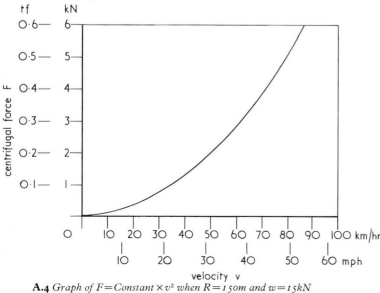

A.4 *Graph of F=Constant × v² when R=150m and w=15kN*

From the graph it is clear that when the velocity is low then a small increase in velocity causes a small increase in centrifugal force but when the velocity is greater then a small increment in velocity makes a progressively larger increase in the centrifugal force.

If a larger radius were used in the calculation then the curve of the graph would become flatter and a particular centrifugal force would correspond to a higher speed. Since the centrifugal force must be kept within an acceptable limit, the scale of the road alignment will become larger as the design speed goes up.

(c) *Relation between R and V when F is constant*
The relation between radius and speed when the centrifugal force is constant has a particular significance in the design of roads because of the relevance of centrifugal force to driver comfort.

As we have seen, when a car is driven round a curve a centrifugal force is exerted on the driver. This force tends to dislodge the driver from a comfortable driving position. The faster the speed for a particular radius, the more discomfort is experienced. In addition the centrifugal force acting on the car makes it more difficult to control. There comes a point when the driver will elect to slow down for his own comfort and safety.

The relationship between curvature, speed and driving comfort is important in the design of flowing alignments for major roads since the radii chosen must be sufficiently large for the design speed to produce satisfactory driving conditions. It is also important in the design of minor roads where it is desirable to limit the maximum speed since the radii can be selected to act as one of the limiting factors (as has been mentioned this could be a consideration with flowing alignments also).

Experimental work to correlate radius of curvature and speed has been carried out in connection with the design of major roads. The importance of this factor in the design of minor roads for low speeds has only recently been appreciated and consequently only a small amount of experimental work has so far been completed. However a research programme is now being undertaken by the Road Research Laboratory on behalf of the Department of Transport. In the meantime designers must use their experience and good sense with some assistance from the figures which are available.

As well as ensuring driver comfort every design must, of course, provide an adequate factor of safety against skidding. Information on this topic is obtainable from textbooks on highway engineering (see the bibliography at the end of the book).

In considering the relationship between radius and velocity we will take into account the effect of super-elevation.

The road surface on a curve is given a crossfall which partly offsets the effect of centrifugal force. The forces acting on a vehicle traversing a curve on a super-elevated surface are shown in **A.5**, in which N is the force normal to the road surface and P is the friction force between the tyres and the road surface.

Resolving forces along the road surface:

$$P = \frac{Mv^2}{R} \cos \theta - W \sin \theta$$

Resolving forces normal to the road surface:

$$N = \frac{Mv^2}{R} \sin \theta + W \cos \theta$$

Driver comfort is normally considered in relation to the side friction factor, f. A value for this factor is determined which, if not exceeded, will ensure comfortable driving conditions.

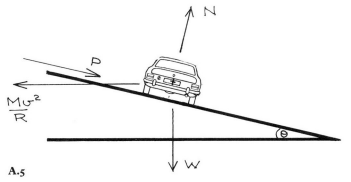

A.5

At design velocity v_D $\quad \dfrac{P}{N} \not\gg f$

therefore

$$\frac{\dfrac{Wv_D{}^2}{gR} \cos \theta - W \sin \theta}{\dfrac{Wv_D{}^2}{gR} \sin \theta + W \cos \theta} \not\gg f$$

ie,

$$\frac{\dfrac{v_D{}^2}{gR} - \tan \theta}{\left(\dfrac{v_D{}^2}{gR}\right) \tan \theta + 1} \not\gg f$$

The value of the expression $\left(\dfrac{v_D{}^2}{gR}\right) \tan \theta$ is small; the denominator can be taken as unity and the equation becomes

$$\frac{v_D{}^2}{gR} - \tan \theta \not\gg f$$

if $\quad \tan \theta = e$

then $\quad \dfrac{v_D{}^2}{gR} \not\gg f + e$ and $R \not\ll \dfrac{v_D{}^2}{g(f + e)}$

The minimum allowable radius is given by

$$R = \frac{v_D{}^2}{g(f + e)}$$

The value of side friction factor recommended by the Department of the Environment for design purposes in *Roads in Urban Areas*[53] is 0.18 for design speeds less than 30mph and 0.15 for speeds in excess of 30mph. The value recommended in *The Layout of Roads in Rural Areas*[54] is 0.15. The maximum value of super-elevation in both these documents is 7%.

Using f = 0.15 and e = 0.07 the graph relating radius and velocity has been drawn, **A.6**. As can be seen, at the higher velocity end of the graph small increases in speed produce relatively large increases in radii. This underlines the fact that the scale of the road alignment increases rapidly at the high speed end of the range. *The Layout of Roads in Rural Areas* lists, for particular design speeds, an absolute

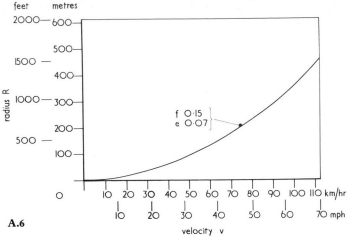

A.6

minimum radius with maximum super-elevation and a minimum desirable radius. For example, at 80km/h the former is given as 230m and the latter as 420m. Normal and minimum radii are listed in *Roads in Urban Areas*.

A way of visualizing the effect of increases in velocity on the minimum radius when $\mu = .15$ and e = .07 is to draw the various radii required for different velocities to the same scale. This has been done in **A.7**.

Another factor which may condition the size of the horizontal curves is the need to provide adequate sight lines. However since our purpose here is simply to underline the relation between speed and the scale and form of the alignment we will only consider this factor in relation to vertical curves. Similar considerations apply to the curves of the horizontal alignment.

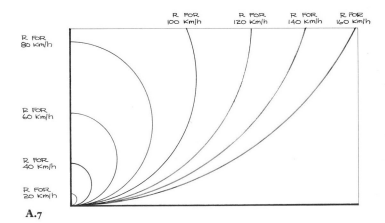

A.7

TRANSITION CURVES

If a straight length of road were connected directly with a circular curve the driver would experience a sudden change of direction as he moved from one to the other. The same would be true if he moved from one circular curve to another. (If the circular curves were in the same direction the impact of the change in direction on the driver could be small, this being a function of the difference in radii of the two curves.)

The purpose of a transition curve is to effect a gradual change in direction at a constant rate and so improve the driving characteristics of the alignment. The need to use transition curves for this purpose becomes more urgent as the design speed increases. However even when transitions are not strictly necessary their incorporation will often improve the appearance of the alignment.

If the driver moves from, say, a straight into a circular curve he can make the transition most easily if it is possible to do so by turning the steering wheel at a constant rate. This imparts a rythmic movement to the act of driving rather than a series of jerking movements. When the steering wheel turns at a constant rate so does the angle which the car wheels make with the axle. Consequently the radius of curvature of the line followed by the car changes at a constant rate also.

The curve which the vehicle follows in this circumstance is a transition spiral or clothoid. In this curve the rate of change of curvature is constant, the radius of curvature being exactly inversely proportional to the length. This is a very beautiful curve and has many manifestations in nature, **A.8**.

The equation of the clothoid, **A.9**, is $\dfrac{1}{r} = \dfrac{l}{RL}$

where

   r = radius of the curve at any point P along its length
   R = the radius at the end of the transition curve, point S
   l = the length measured along the curve from the point of origin, O, to the point P
   L = the total length of the curve, OS

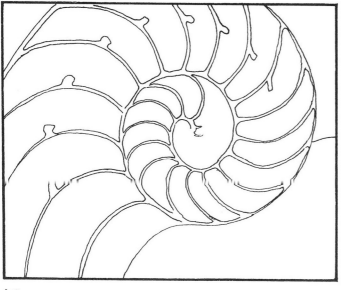

A.8

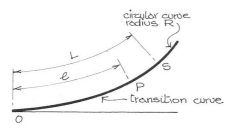

A.9

If a vehicle is driven along this curve at constant speed the rate of change of centrifugal force will be constant.

In practise the curves which are usually adopted only approximate to the clothoid and are chosen for ease of calculation and setting out. In this country the lemniscate is often used. The rate of change of radius with this curve is slightly less than with a spiral. When driving along a lemniscate transition the rate of turning into the curve decreases slightly as the minimum radius is approached.

The lemniscate is a symmetrical curve which can be used with large deflection angles, an extreme case being when the exterior deflection angle between the tangents is 270°. In this case, the curve doubles back on itself and crosses the original starting point at right angles to its original direction, thus forming one loop of the figure eight.

The length of spiral transition should be chosen to meet the requirements of drivers in two respects. They should be able to traverse the curve by turning the steering wheel at a rate which they find convenient and when they drive round the curve at the design speed the rate of increase of centrifugal force should be one which they find comfortable. A number of different authorities have suggested ways of determining the lengths of spirals to meet these requirements.

THE STRAIGHT ALIGNMENT

From the dynamic point of view the straight alignment would appear, at first examination, to be ideal. No uncomfortable centrifugal force operates on the driver and distraction from changing views are minimized, since attention tends to be focused on the route ahead. However the disadvantages of straight lengths are considerable. On wide roads especially it is difficult for the driver to gauge his speed and the monotony of long straights can lead to boredom and inattention.

The absence of centrifugal force operating on the driver's body and the vehicle means that he has difficulty in gauging the amount of control which he exerts over his machine. We have mentioned in another section that curves are preferred in nature to straights and one of the reasons for this is that motion in a curved line is more easily controlled than that on one which is straight. It is true that acceleration in a straight line causes a force to be exerted on the drivers body but since he is pressed back into his seat this does not impart any sense of insecurity and the same pressure is experienced for the same acceleration rate whether he is travelling at a low or a high speed. On a straight, smooth-surfaced road, apart from his view of the surroundings the main impression of high speed is conveyed by the difficulty the driver has in controlling the vehicle.

Another factor which militates against the use of straights is the infrequent occurrence of very large flat areas in the British landscape. Curved lines usually fit our topography best whereas long straights can lead to unsightly earthworks and scars in the ground formations. In minor roads, where the adoption of low speeds is required, the existence of long straights encourage an opposite trend.

Nevertheless straight lengths of road are quite often required and can be satisfactory provided they are not too long and are properly integrated with the other elements of the alignment. In the residential urban and suburban

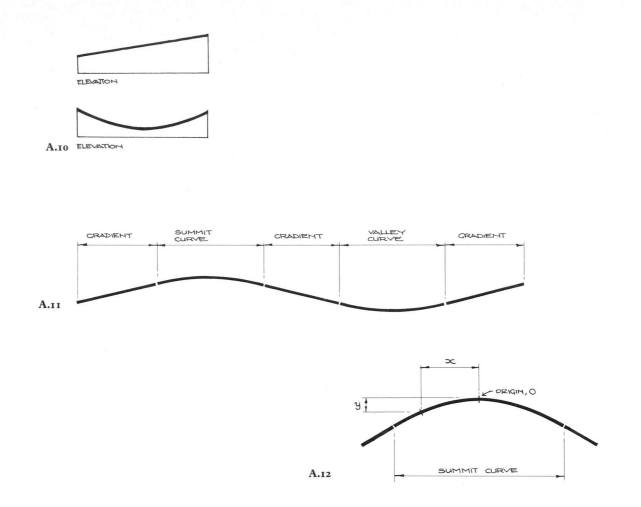

ELEVATION

A.10 ELEVATION

GRADIENT   SUMMIT CURVE   GRADIENT   VALLEY CURVE   GRADIENT

A.11

ORIGIN, O

SUMMIT CURVE

A.12

context for example there is a long tradition of short straight lengths of road being used successfully in conjunction with parallel terraces.

A need for straight lengths is likely to arise on roads without a central reserve between lanes with traffic in opposite directions. A curved line may not permit the long passing sight lines required.

2 The vertical plane
The two elements of the vertical alignment are gradients and vertical curves, **A.10**. Vertical curves may be summit curves or valley curves, **A.11**.

GRADIENTS
The gradient is defined as the number of metres of vertical rise or fall in 100m horizontal distance. If the ground rises 5m in 100m this is called a 5% gradient.

The minimum value is determined by considerations of drainage and is usually taken to be about 0.4% (1:250). The maximum is a function of the road classification and the design speed. The normal maximum is 4% (1:25) for other than access roads and for access roads a maximum gradient of 7% (1:14) is usually considered satisfactory.

The designer should bear in mind the tendency to move more quickly downhill than uphill and the possibility in some circumstances of using gradients to modify vehicle speed.

VERTICAL CURVES
The configuration of vertical curves is determined by the need to preserve smooth driving conditions and to maintain minimum sight lines. The appearance of the road alignment should also be a major consideration in their design. This last aspect is not always given the consideration it warrants.

The curve used is the parabola. When properly proportioned it provides a smooth alignment and the necessary visibility. In practice if the sight line requirements are met then the dynamic characteristics will usually be adequate. By this it is meant that no sense of discomfort will be experienced by the driver due to the centrifugal force generated by the vertical curve provided the speed of the vehicle is kept within the design limits. However by meeting the sight line requirements it does not follow that a visually pleasing alignment will result and this must be given special consideration.

It is often the case that in order to preserve a visually satisfactory alignment it is necessary to supply sight lines which are longer than the minimum, and the aesthetics of alignment rather than purely engineering considerations becoming the determining criterion.

The parabola is a particularly convenient curve to use because of the ease with which the coordinates can be calculated. It is mainly for this reason that it is selected in preference to a circular curve or a spiral transition.

In the parabola, **A.12**, the y coordinate is equal to the square of the x coordinate multiplied by a constant, so the equation is $y = kx^2$

The rate of change of slope of a parabola is a constant

$$\frac{d^2y}{dx^2} = 2k$$

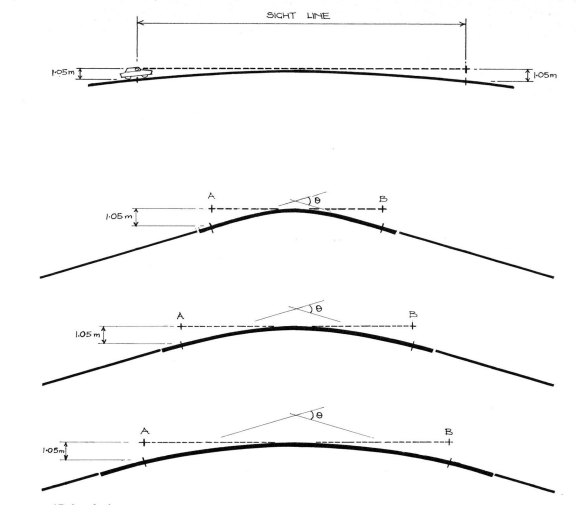

SIGHT LINE

1·05m   1·05m

A.13

A   θ   B

1·05 m

A   θ   B

1.05 m

A   θ   B

1·05m

A.14 *θ is constant; AB, length of sight line, increases with the design speed; the length of the vertical curve increases with the design speed*

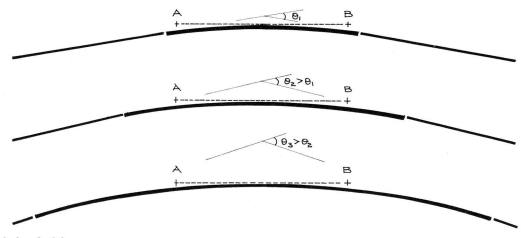

A θ_1 B

A $\theta_2 > \theta_1$ B

A $\theta_3 > \theta_2$ B

A.15 *θ varies; the length of the sight line, AB, is constant; the length of the vertical curve increase with increase in θ*

Thus the driving characteristics of this curve are similar to those of a circular curve, since the rate of change of slope would be virtually constant in a circular curve for the size of radius and length of curve which would be used.

In order to obtain a satisfactory length for the vertical curve it must be considered in relation to the horizontal aspects of the road line. This topic of the integration of the vertical and the horizontal alignment is considered in Chapter 5, Internal Harmony.

In the case of summit curves the length of the curve must be at least sufficient to provide the necessary sight lines. The driver must be able to see far enough over the crest of a hill, from any point on the approach path, so that he will be able to stop before reaching a stationary obstacle which may come into view or, on two lane roads, so that he will be able to overtake safely. In measuring the sight line it is assumed that the driver's eye level is 1.05 m above the road surface and that he is viewing a point which is also located 1.05 m above road level, A.13. (For residential roads, in situations where visibility is required between driver and young children a 600 mm height is recommended at the position being viewed.)

The design speed and the difference in angle between the tangents at the beginning and end of the vertical curve, θ, is critical in determining the required length of the vertical curve to supply adequate sight lines.

For any particular difference in approach gradients or tangents, the greater the design speed the longer must be the vertical curve, A.14. Also when the design speed is constant then the larger the difference in angle between the tangents the longer must be the vertical curve, A.15.

Determination of the minimum length of sight line is a crucial factor in the design of the vertical curve. Although, as has been pointed out, it is often not adequate to use the minimum length because of the aesthetic considerations it is nonetheless vital to ensure that at least a safe minimum is being provided. The length of the sight line is a function of the design speed since the faster the vehicle is travelling the longer distance it will require in which to stop or overtake. On dual carriageways the criteria is the stopping distance. This is made up of two components; the distance the vehicle travels in the time between the driver deciding to apply the brakes and the brakes coming into effect (called the reaction time) and the distance travelled by the vehicle after the brakes are in operation (called the braking distance). The reaction time and the braking distance have been determined experimentally and the resulting stopping distances for design purposes are listed in the relevant Department of the Environment publications. These figures are reproduced in Table 10.

On single carriageway roads minimum stopping visibility distance must be provided but in addition it is necessary to incorporate the minimum overtaking visibility distance over as great a proportion of the road alignment as possible. Overtaking visibility distances are much longer than stopping distances and so larger radius curves and more frequent tangent lengths are required. The Department of the Environment recommendations for minimum overtaking visibility distances are listed in Table 11.

TABLE 10 Minimum stopping distances

Design speed		Minimum stopping distance	Sources
mph	km/h	m	
70	120	300	A
60	100	210	A
50	80	140	A
40	60	90	A
30	48	60	B
25	40	45	B
20	32	33	B
15	24	23	B
10	16	14	B
5	8	6	B

Sources:
A *Department of the Environment*, The Layout of Roads in Rural Areas, HMSO, London, 1968
B *Department of the Environment*, Residential roads and footpaths: layout considerations, *Design Bulletin 32*, HMSO, London, 1977

TABLE 11 Minimum overtaking distances

Design speed		Minimum stopping distance	Sources
mph	km/h	m	
60	100	450	A
50	80	360	C
40	60	270	C
30	50	225	C
20	30	135	C

Sources:
A *See Table 10*
C *Department of the Environment*, Residential Roads and Footpaths: layout considerations, *Design Bulletin 32*, HMSO, London, 1977

The physical criteria which govern the minimum length of a valley curve are comfort and headlight distance. A method of calculating the minimum curve lengths of both summit and valley curves is presented by the Department of the Environment.

The aesthetic considerations including the desirability of having the vertical and horizontal aspects of the alignment 'in-phase' may well be the overall governing criteria for both types of curve, requiring longer lengths than the other considerations.

APPENDIX B
ROAD LAYOUT DATA

Contents

In Appendix A we have illustrated the relationship between speed and the scale of the road alignment. Elsewhere we have emphasized that flowing alignment is a function of the dynamics of vehicles moving at high speeds. When low speeds are called for the type of alignment designed should be quite different.

There has been general agreement for many years now regarding the appropriate data for the design of flowing alignments. However, in the case of alignments for low speeds, recent years have seen a revision in approach. The new consensus in relation to residential areas is embodied in the Department of the Environment's Design Bulletin 32.[55] Their recommendations and information relating to the design and layout of urban roads are now contained in Design Bulletin 32 (residential roads) and in *Roads in Urban Areas*[56] (all other urban roads).

The main purposes of this appendix are to reproduce some of the data from these publications so that the reader may contrast the data and criteria for low speed road design with

those appropriate for the design of high speed roads and to present some material for preliminary reference. Local distributors, which are treated in *Roads in Urban Areas*, and access roads, which are designed with the revised approach, are considered here and they are put in context by a preliminary outline of the road hierarchy.

In preparing a particular road design the relevant document of the two listed above must be studied.

A section is included on the location of services, which can affect access road design particularly.

1 The road hierarchy

A town is conceived, in present day planning practice, as being made up of a number of interconnected districts. These are more or less distinct areas which may be mainly residential, or devoted to business or industry. Alternatively they can be areas of mixed development but with certain features that render them distinct entities in the total conurbation.

Each district is held to consist of a number of environmental areas in which environmental considerations take precedence over the demands of traffic and through traffic is excluded or discouraged. Each environmental area is served by a system of local roads which connect to a network of local distributors. These in turn link up with other roads between districts which finally lead into the trunk routes connecting the urban areas. The whole road system forms a hierarchical network which is represented diagrammatically in **B.1**.

Urban roads then can be divided into four main categories. These are: primary distributors, district distributors, local distributors and access roads. Each category has its own particular function:

Primary distributors form the primary network for the town as a whole and all longer-distance traffic movements to, from and within the town are canalised on to such roads.

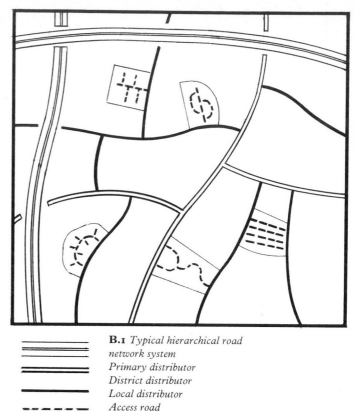

B.1 *Typical hierarchical road network system*
Primary distributor
District distributor
Local distributor
Access road

District distributors distribute traffic between the residential, industrial and principal business districts of the town and form the link between the primary network and the roads within residential areas.

Local distributors distribute traffic within districts. In residential areas, they form the link between district distributors and residential roads and should not normally give direct access to dwellings.

Residential access roads link dwellings and their associated parking areas and common open spaces to distributors.

TABLE 12 Design flows of two-way urban roads

Road type	Two-lane carriageway					Undivided carriageway				Dual carriageway		
	Peak hourly flow vph, both directions of flow†					Peak hourly flow vph, one direction of flow				Peak hourly flow vph, one direction of flow		
						4 lane			6 lane	Dual 2 lane		Dual 3 lane
	6.1m	6.75m	7.3m	9m	10m	12.3m	13.5m	14.6m	18m	Dual 6.75m	Dual 7.3m	Dual 11m
Urban motorway											3600	5700
All purpose road, no frontage access, no standing vehicles, negligible cross traffic			2000		3000	2550	2800	3050		2950*	3200*	4800*
All purpose road, frontage development, side roads, pedestrian crossings, bus stops, waiting restrictions throughout day, loading restrictions at peak hours	1100	1400	1700	2200	2500	1700	1900	2100	2700			

Notes:

† *60/40 directional split can be assumed*

* *Includes division by line of refuges as well as central reservation; effective carriageway width excluding refuge width is used*

2 Local distributors

The data, diagrams and tables in this section are reproduced from the following Department of the Environment publications: *Roads in Urban Areas*, Technical Memorandum H9/76 entitled *Design Flows for Urban Roads*, and Technical Memorandum H11/76 entitled *The Design of Major/Minor Priority Junctions*, by permission of the Controller of Her Majesty's Stationery Office.

ROAD CAPACITY

The capacity of local distributor roads is based on tables 12, 13 and 14, reproduced from Technical Memorandum H9/76. This memorandum should be studied in conjunction with the tables.

TABLE 13 Design flows on one way urban roads

Road type	Carriageway width and vph, one direction of flow					
	6.1m	6.75m	7.3m	9m	10m	11m
All purpose road, no frontage access, no standing vehicles, negligible cross traffic		2950	3200			4800
All purpose road, frontage development, side roads, pedestrian crossings, bus stops, waiting restrictions throughout day, loading restrictions at peak hours	1800	2000	2200	2850	3250	3550

The recommended flows allow for a proportion of heavy vehicles equal to 15%. No allowance will need to be made for lower proportions of heavy vehicles; the peak hourly flows at the year under consideration should be reduced when the expected proportion exceeds 15%, as set out in Table 14.

TABLE 14 Heavy vehicle content

Heavy vehicle content	Total reduction in flow level (vph)		
	Motorway and dual carriageway all purpose road	10m wide and above single carriageway road	Below 10m wide single carriageway road
	per lane	per carriageway	per carriageway
15–20%	100	150	100
20–25%	150	225	150

DESIGN SPEED

This is normally 30mph (50km/h); see *Roads in Urban Areas*, section 3.2.

SIGHT DISTANCES

These must be measured 1.05m above the carriageway, for vertical and horizontal alignment, along the centre lines of both the near and offside lanes of the carriageway.
Minimum sight distances for various design speeds are given in Table 15 below, which reproduces Table 3.1 of *Roads in Urban Areas*.

TABLE 15 Minimum sight distances

Design speed	Sight distances	
	Minimum overtaking distance (single carriageway)	Minimum stopping distance (single and dual carriageways)
km/h	m	m
80	360	140
60	270	90
50	225	70
30	135	30

VERTICAL CURVES

The minimum length should be either
1) that from the formula $L = KA$
    where L = curve length in metres
        A = algebraic difference in gradients (expressed as a percentage)
        K = value from Table 16
or
2) the length shown in the fourth column of Table 16 if this gives a larger value than (1).
Table 16 reproduces table 3.3 from *Roads in Urban Areas*.

TABLE 16 Minimum vertical curve lengths

Design speed	Minimum K value for overtaking	Minimum K value for stopping and comfort	Minimum vertical curve length
km/h			m
80	—*	25	50
60	—*	10	40
50	60	6	30
30	20	1	20

Values not quoted as dual-carriageway layouts will normally be appropriate for these design speeds

LANE WIDTH

7.3m—Single two lane carriageway in industrial districts
6.75m—Single two lane carriageway in principal business districts
6.00m—Minimum single two lane carriageway in residential districts used by heavy vehicles.

BENDS

Minimum radius for a design speed of 30mph (assuming 4% super-elevation) is 90m but 200m is considered normal.
Minimum radius for a design speed of 30mph (assuming 7% super-elevation) is 80m but 115m is considered normal.
Transition curves are required if radius is less than 300m. The use of transitions with larger radii is often advisable to improve the appearance and assist in the introduction of super-elevation.

TABLE 17 Required widening on bends to allow for vehicle overhang

Centre line radius before widening	Total carriageway width required on carriageways having a normal width of:		
	6m	6.75m	7.3m
Less than 150m	7.2	7.9	7.9
150–300m	7.0	7.3	7.3
300–400m	6.6	7.3	7.3
More than 400m	6.0	6.75	7.3

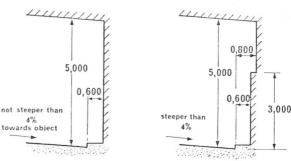

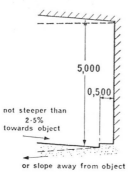

B.2 *Recommended minimum clearances for 50km/h*

CROSSFALL

Except on curves where super-elevation or elimination of adverse crossfall or camber is required crossfall should not normally be steeper than 1 in 40 nor flatter than 1 in 48 from the crown downwards towards the side of the road.

CLEARANCE

Recommended minimum clearances are given in Table 18 and figure **B.2**, reproduced from *Roads in Urban Areas.*

TABLE 18 Horizontal clearances from the carriageway

Design speed km/h	Height of object on footway, verge or central reserve	Minimum clearance where carriageway crossfall is:		
		away from object, or towards object but not steeper than 2.5%	towards object but not steeper than 4%	towards object and steeper than 4%
50	Less than 3m	0.500m	0.600m	0.600m
	3m and above	0.500m	0.600m	0.800m
60	Less than 3m	Minimum in all cases		0.600m
		Desirable where conditions permit		1.250m
	3m and above	Minimum in all cases		1m
		Desirable where conditions permit		1.250m
80	All heights	Minimum in all cases		1m
		Desirable where conditions permit		1.500m

FOOTWAYS

In principal business and industrial areas the recommended width is 3m and in residential districts 2m. If no footway is required a 1m minimum width verge should be provided.

TYPICAL CROSS-SECTION

Recommended single two-lane carriageway widths are 7.3m in industrial districts, 6.75m in principal business districts and 6m minimum in residential districts used by heavy vehicles.

Footway widths should be 3m in principal business and industrial districts and 2m minimum in residential districts.

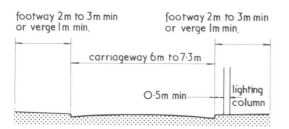

B.3

Verges should be a minimum of 1 m if provided instead of a footway, but consideration should be given to the width needed for traffic signs, underground services, etc, and clearances in accordance with Table 18, above.

JUNCTION SPACING

90m minimum is suggested as a rough guide. Where a staggered cross roads between an access road and a local distributor is necessary (a four-legged junction) the minimum recommended spacing between legs of the access road is 40m, **B.4**.

VISIBILITY AT JUNCTIONS

Visibility should be measured between points 1.05 m above road level. Some reduction in the 9m distances may be reasonable for lightly trafficked minor roads (not housing estate roads). On such roads the 9m distance may be reduced to 4.5m in difficult circumstances, but the 90m distance should remain the same, **B.5**.

If the major road is one-way a single visibility splay in the direction of approaching traffic will suffice. If the side road serves as a one-way exit from the major road, no visibility splay will be required provided forward visibility for turning vehicles is adequate.

CORNER RADII

A kerb radius of 10.5m will normally suffice in urban areas for junctions used by commercial vehicles. A radius of 6m will be adequate for those junctions used by smaller vehicles. In rural areas a value of 25m is preferred.

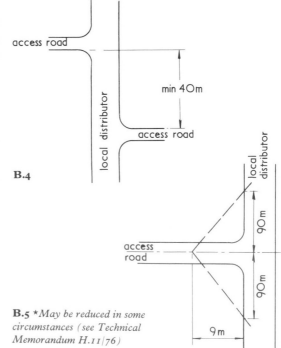

B.4

B.5 **May be reduced in some circumstances (see Technical Memorandum H.11/76)*

3 Residential roads

The data, diagrams and tables in this section are reproduced from Design Bulletin 32, *Residential Roads and Footpaths: layout considerations*, by permission of the Controller of Her Majesty's Stationery Office. Where noted diagrams have been drawn from data in Design Bulletin 32.

VEHICLE SIZES

In the preparation of Design Bulletin 32 the Department of the Environment used certain sizes for the various categories of vehicle; data regarding these sizes is reproduced below under each section.

Private cars

The vehicle was chosen as a result of a statistical analysis of the turning circles and wheelbases of private cars registered in 1971. The relevant graphs are reproduced in **B.6**, **B.7** and **B.8**.

The vehicle chosen for use in design was the BLMC 1800 Mk 11, this being representative of the 95th percentile car. Details are given in **B.9** and Table 19.

TABLE 19 BLMC 1800 Mk II with extended front overhang

Condition	Lock	Turning circle diameter	
		m	in
Between	left	11.24	443
kerbs	right	11.56	455
Between	left	12.06	475
walls	right	12.28	484

Fire fighting vehicles

Details of the vehicle selected are presented in **B.10** and Table 20. It should be emphasized that the vehicle is of the type used primarily for dwellings with floor levels below 9 m and that for multi-storey dwellings larger vehicles with hydraulic platforms may be required and it is essential that fire authorities be consulted with regard to their needs. (See also *Access for Fire Appliances*, Fire Prevention Note 1/70.[57])

TABLE 20 Fire engine—Dennis with simulated wheeled escape Chassis No 3897 F15 Body No 44647

Condition	Lock	Turning circle diameter	
		m	in
Between	left	15.82	623
kerbs	right	15.43	607
Between	left	17.52	690
walls	right	17.25	679

Refuse collection vehicles

The characteristics of nineteen vehicles were examined on the basis of details obtained from three leading manufacturers and the local borough depot. The vehicle amongst these with the largest between-walls turning circle was selected as representative, **B.15**. A comparison between the sizes for this vehicle and the others is shown in **B.11** to **B.14**. The main dimensions for the representative vehicle are shown in **B.15** and Table 21.

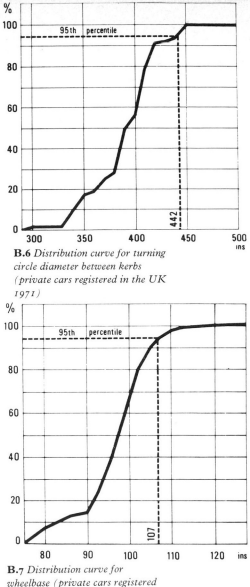

B.6 *Distribution curve for turning circle diameter between kerbs (private cars registered in the UK 1971)*

B.7 *Distribution curve for wheelbase (private cars registered in the UK 1971)*

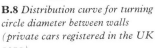
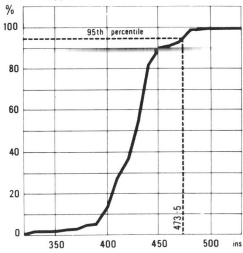

B.8 *Distribution curve for turning circle diameter between walls (private cars registered in the UK 1971)*

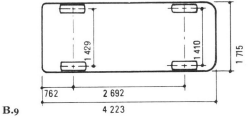

B.9

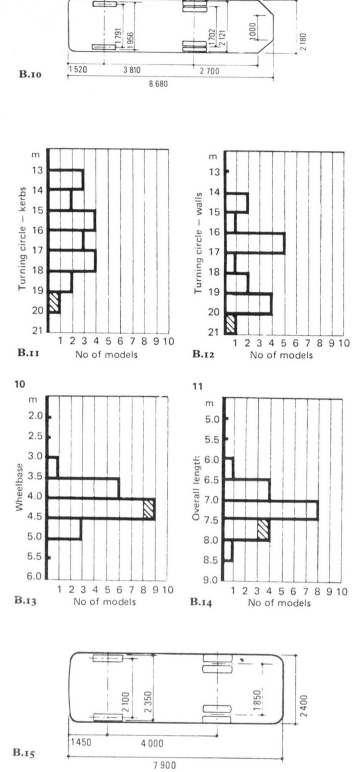

B.10

B.11 No of models

B.12 No of models

B.13 No of models

B.14 No of models

B.15

TABLE 21 Refuse vehicle—Shelvoke Drewry Pakamatic type TBY (Leyland engine and transmission) Unladen WT 9050 kg (19936 lb) G.V.W.: 14170 kg (32700 lb)

Condition	Lock	Turning circle diameter	
		m	in
Between kerbs	left	19.25	758
	right	17.20	677
Between walls	left	20.25	797
	right	18.10	713

Furniture removal vehicles (pantechnicon)

Many removal firms consider pantechnicons as coming in three basic functional sizes: small vehicles with a capacity of 700–1000 ft³, medium vehicles with a capacity of 1300–1600 ft³ and large vehicles with a capacity of 1800–2200 ft³. The great majority of domestic removal work is done by the small and medium sized vehicles.

Vehicles are rarely built up to the legal maximum length of 11 m (36 ft 1 in) the largest being between 8.5 m (about 28 ft) and 10 m (about 33 ft).

The majority of vehicles, the medium capacity vans, are built on chasses of wheelbase between 150 in and 175 in. The larger capacity vehicles start at 180 in and go up to the extreme case found of 250 in.

The most commonly used chassis for Luton-bodied furniture vans (those that have a separate cab with the van body cantilevered over the top of it) are from the Bedford K range and the Ford D range, **B.16**. For the true pantechnicon bodies (those that have an integral cab and body) bus chasses are probably the most common. Bus chasses are popular because of their good manoeuvrability, usually having smaller turning circles than lorry chasses. Details of the vehicle chosen for tests are shown in **B.16** and Table 22.

TABLE 22 Luton Marsdens' body on Ford D800 C.V. chassis Capacity: 56.6 m³ (2000 ft³) Unladen WT: 5296 kg (11676 lb) Height: 3960 mm (156 ins) Gross plated WT: 13163 kg (29100 lb)

Condition	Lock	Turning circle diameter	
		m	in
Between kerbs	left	18.07	711
	right	20.90	823
Between walls	left	19.40	764
	right	21.92	863

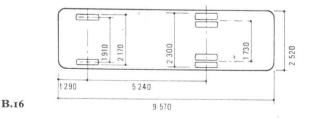

B.16

STOPPING DISTANCES

Figure **B.17** gives a range of stopping distances commensurate with various vehicle speeds. These are based on the reaction times and deceleration rates implied in Table 3.1 in *Roads in Urban Areas*.

The distances are intended to cater for the majority of vehicles and drivers in most weather conditions and may therefore safely be used as general guidance in the design of residential access roads.

SERVICE VEHICLES: ACCESS TO DWELLINGS

The layout should be so designed that at least the distances shown in Table 23 are obtained (see p. 292).

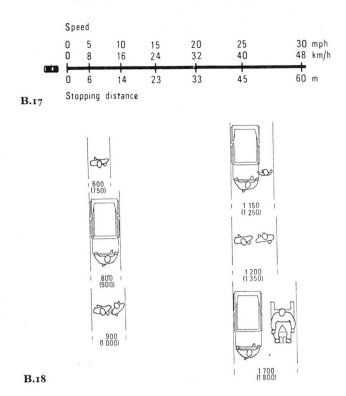

Speed
0	5	10	15	20	25	30 mph
0	8	16	24	32	40	48 km/h
0	6	14	23	33	45	60 m

B.17 Stopping distance

600 (750)

1 150 (1 250)

80'0 (900)

1 200 (1 350)

900 (1 000)

1 700 (1 800)

B.18

DESIGN SPEED

Design Bulletin 32[58] does not specify speeds for roads in residential areas. It says, however, that in order to provide a safe environment for pedestrian movement, it is necessary in design to ensure that 'the overall configuration, geometry and edge conditions alongside the carriageways encourage drivers to keep to speeds of well below 30mph (48km/h)'.

FOOTWAY AND FOOTPATH WIDTHS

The spaces required for various types of pedestrian movement are shown in **B.18**. The diagrams illustrate the minimum spaces necessary. For greater ease of movement, especially over long distances, the dimensions shown in brackets should normally be used.

ROAD WIDTHS

The road width used is a function of vehicle width, tolerances required and the needs of cyclists. The tolerances allowed by some carriageway widths for vehicles already described are shown in **B.19**, reproduced with the accompanying notes from the Department of the Environment's Design Bulletin 32.

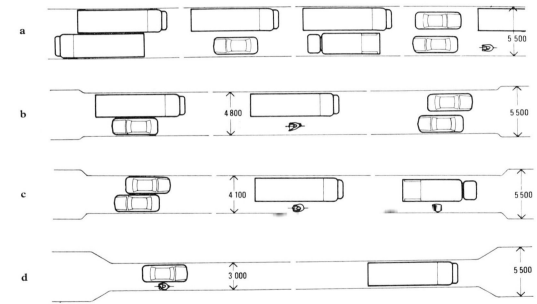

a — 5 500

b — 4 800 — 5 500

c — 4 100 — 5 500

d — 3 000 — 5 500

B.19

(a) A 5.5 m width allows all vehicles to pass each other, with an overall tolerance of 0.5 m for the largest vehicles but with ample clearance for all others. Given the infrequency of large vehicles on residential roads, this width will normally be the maximum required to cope with residential traffic. Below 5.5 m the carriageway will be too narrow for the free movement of pantechnicons. Where such vehicles are allowed access passing places may be required. The carriageway width required between passing places will then depend upon the combinations of vehicle types expected; the frequency with which vehicles may meet each other and the delay which may be caused to traffic movement. These factors may be expected to vary with traffic volume.

(b) At 4.8 m the carriageway will allow a wide car to pass a pantechnicon with an overall tolerance of 0.5 m, and traffic may therefore still be regarded as being in free flow.

(c) At 4.1 m the carriageway will be too narrow for pantechnicons to pass vehicles other than cyclists. It does, however, allow wide cars to pass each other with an overall tolerance of 0.5 m.

Hence, while being more restrictive on the movement of large vehicles, a width of 4.1 m will still provide two-way flow for the majority of residential traffic.
Below 4 m the carriageway will be too narrow for private cars comfortably to pass each other except at very low speed and may be uncomfortable for cyclists in conjunction with large vehicles. Widths of less than 4 m therefore should be regarded as catering only for single-file traffic.

(d) The choice of width below 4 m will depend largely upon the frequency and ease with which cyclists and cars may need to pass each other. It is recommended that 3 m be regarded as the minimum width between passing bays on a single-track system e.g. figure d, but that where narrow sections are introduced solely as pinch points, e.g. figure c, a width of 2.75 m will normally be adequate to allow all vehicles to pass through on their own. Widths below 2.75 m may, however, be considered where private cars only are allowed access, for example in private drives. In such circumstances clearances may be necessary to allow passengers and drivers to get out of the vehicle at rest.

Parking

On-street parking causes accidents. The Department of the Environment's Design Bulletin 32 states that: 'In almost half of all pedestrian accidents and a quarter of vehicular accidents in residential areas the presence of a parked vehicle was reported. Stationary vehicles can cause hazards by masking pedestrians and by masking moving vehicles from each other. The aim therefore should be to minimise the use of carriageways for parking'.

It continues: 'For carriageways to be kept clear of parked cars it is necessary in design to ensure that:

(a) off-street parking spaces are provided to suit the needs of residents, visitors, service and maintenance vehicles for both short term and long-stay parking;

(b) routes between parking spaces and dwelling entrances or other destinations are shorter and more convenient to use than would be the case if parking were on the carriageways;

(c) parking spaces for use by service vehicles are not occupied at the times they are required by residents or visitors' cars, and vice-versa.

. . . on-street parking is unlikely to occur on roads which do not give direct access to dwellings, and in such situations the carriageway width may largely be determined by considerations of moving traffic.

An example of this is shown in [B.20]. Similar conditions may be created where adequate off-street parking is provided between the carriageway and the dwelling entrance and/or where the off-street facility offers a more convenient parking location than the carriageway.

The extent to which on-street parking occurs will also be affected by the width of the carriageway itself, and it may be possible to use narrowed carriageways to discourage on-street parking and to use variations in carriageway width to control where parking will occur. For example at the intersection with a pedestrian route a narrow carriageway may be used not only to reduce the distance across the road but also to ensure good visibility between pedestrians and moving vehicles by discouraging parking at the crossing point [B.21].'

Emergency access

Carriageways as narrow as 2.75 m may be acceptable to the fire service, provided that roads can be kept clear of parked cars. In all cases it is essential that early discussions are held with the local Fire Authority.

SIGHT LINES

The driver's line of vision in the vertical plane is illustrated in B.22 and its application to vertical and horizontal curves is shown in B.23 and B.24 (which are based on data given in Design Bulletin 32). If measures have been taken to keep the speed of most vehicles well below 30 mph it may be possible to base visibility on these lower speeds.

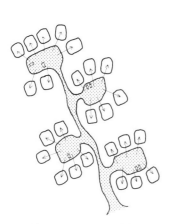

B.20 Where on-street parking is unlikely to occur, narrowed carriageways may be used to reduce the scale of the road

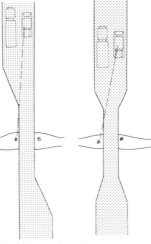

B.21 Narrowed sections of carriageway may be used to discourage parking in situations where pedestrians could be masked from approaching vehicles

B.22 Driver's line of vision in vertical plane

B.23 Sight line on vertical curve

1·05 m ABSOLUTE MAXIMUM 600 mm WHEN VISIBILITY IS REQUIRED BETWEEN DRIVERS AND YOUNG CHILDREN PEDESTRIANS

STOPPING DISTANCE

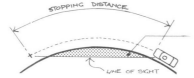

1·05 m ABSOLUTE MAXIMUM HEIGHT OF POTENTIAL OBSTRUCTION TO VISIBILITY IN THIS ZONE. 600 m HEIGHT IF VISIBILITY IS REQUIRED BETWEEN DRIVERS AND YOUNG CHILDREN.

STOPPING DISTANCE

B.24 Sight line on horizontal curve

LINE OF SIGHT

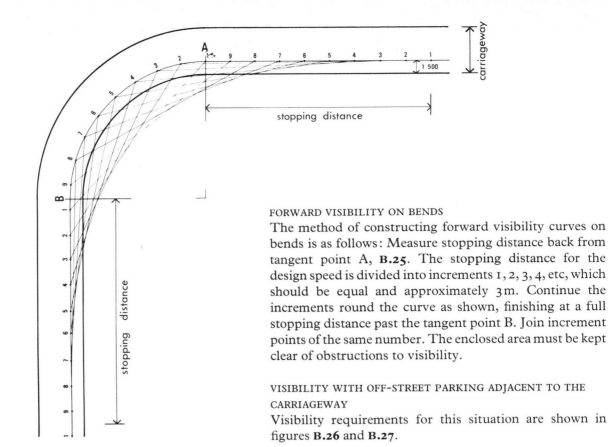

B.25 *Determining forward visibility on bends*

FORWARD VISIBILITY ON BENDS

The method of constructing forward visibility curves on bends is as follows: Measure stopping distance back from tangent point A, **B.25**. The stopping distance for the design speed is divided into increments 1, 2, 3, 4, etc, which should be equal and approximately 3 m. Continue the increments round the curve as shown, finishing at a full stopping distance past the tangent point B. Join increment points of the same number. The enclosed area must be kept clear of obstructions to visibility.

VISIBILITY WITH OFF-STREET PARKING ADJACENT TO THE CARRIAGEWAY

Visibility requirements for this situation are shown in figures **B.26** and **B.27**.

TABLE 23 Maximum distances of service vehicles from dwelling

Type of vehicle	Maximum distance
Refuse collecting vehicle to communal container	9 m
Refuse collecting vehicle: recommended maximum dustbin-carrying distance	25 m
Fire appliances to front or back door of single family dwelling with floors not higher than 6 m above the ground	45 m

Notes:
1 *Refuse vehicle information from CP 306 Part 1 : 1972*
2 *Fire appliance information from Home Office Fire Department Fire Prevention Note N 1/70 and CP 3 Chapter 4 Part 1 : 1971*
  Local fire authorities should be consulted at all times.

VISIBILITY AT JUNCTIONS

To enable drivers emerging from the stem road to see and be seen by drivers proceeding along the through road, unobstructed visibility will be required within the shaded areas of figures **B.28** and **B.29**.

In **B.28**, which shows an access road junction with another access road, the minimum value of Y is the stopping distance at the expected speed of vehicles on the through road.

In **B.29**, which shows an access road junction with a local distributor, Y is based on the time taken for vehicles to turn out from the stem road and on the distance travelled in that time by vehicles on the through road proceeding at a constant speed.

On local distributors where speeds up to 30 mph (48 km/h) may be expected, a Y distance of up to 90 m will usually be required.

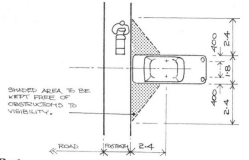

B.26

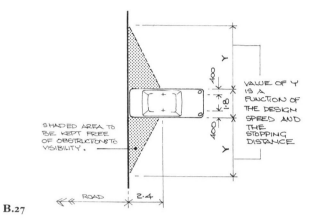

B.27

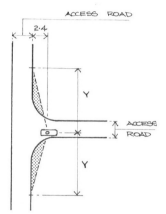

B.28 *Access road junction with access road—for use with low traffic flows*

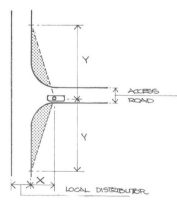

B.29 *Access road junction with local distributor*

For X value, 4.5 m should be sufficient in most cases where traffic volumes do not exceed 300 vph.

GRADIENTS

Steep gradients may be required in some circumstances but due consideration should be given to their effect on speed, especially on bends and at junctions, and to the effect of icy conditions. The normal maximum gradient is 10% (1:10); the preferred maximum gradient is 7% (1:14); at approaches to junctions, including junctions between private drives and public highways, the maximum should be 5% (1:20).

JUNCTION SPACING

For the spacing of access road junctions along local distributor roads see the section above on 'Local Distributors'.

The spacing of access road junctions along access roads should be made with due regard to the economic use of land.

Daylight, sunlight and privacy requirements normally dictate at least 30–40m between junctions, but closer spacings may be advantageous in some places.

SPACING OF PASSING PLACES ON SINGLE LANE, TWO-WAY ROADS

Appendix 4 of the Department of the Environment's Design Bulletin 32 contains a summary of studies undertaken at the Transport and Road Research Laboratory on behalf of the Housing Development Directorate into the use of single-lane carriageways with passing places.

The results suggest that for two-way flows of up to 100 vehicles per hour two passing places within 180m (ie, a separation of 60m, centre-to-centre) are sufficient. For higher flows of up to a two-way total of 300 vph three passing places are probably necessary (separation of 45m, centre-to-centre).

Figures **B.30** to **B.34** are reproduced from Appendix 5 of Design Bulletin 32. 'Considerations for the provision of forward visibility between passing places on narrowed carriageways'; **B.30–B.32** illustrate that the forward visibility distance is determined by the combined stopping distances of opposing vehicles plus the distances required between bays necessary to cope with the traffic volumes envisaged.

In order to minimize the area affected by visibility lines, passing places should normally be supplied on bends, **B.32**, at junctions between narrowed carriageways, **B.33**, and at their entry point from carriageways of 'normal' width, **B.34**.

RELATION BETWEEN HORIZONTAL CURVATURE AND VEHICLE SPEED

Horizontal curvature affects vehicle speed. The Department of the Environment is investigating the relationship, and preliminary results for residential areas are shown in **B.35**. This data should be used only as guidance to the likely speeds on bends of 90° or tighter. Figure **B.35** represents mean velocities and the Department of the Environment recommends an additional safety factor of 20% to be added to the speed when considering the stopping distances and the forward visibility which should be provided.

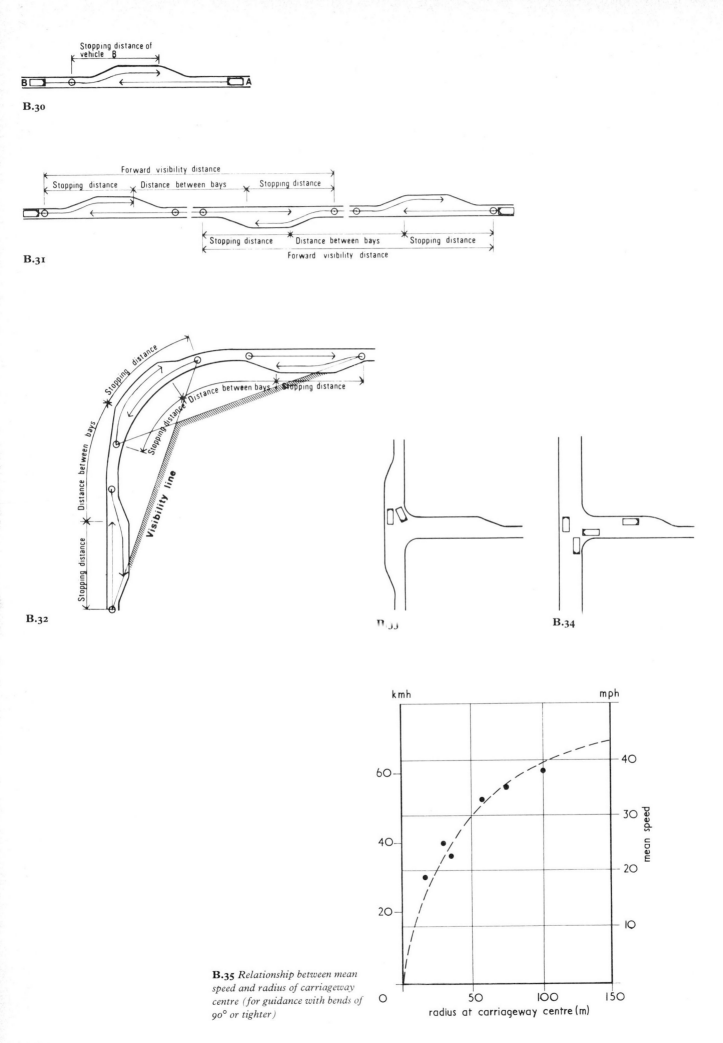

B.30

B.31

B.32

B.33

B.34

B.35 *Relationship between mean speed and radius of carriageway centre (for guidance with bends of 90° or tighter)*

4 Services under roads and footpaths

SCOPE

The services which usually have to be accommodated are electricity, gas, water, telephone, public lighting and sewers, which may be either separate surface water and foul or combined.

RECOMMENDED PRACTICE

This is contained in the report of the Joint Committee on the Location of Underground Services[59] first published by the Institution of Civil Engineers in 1946 and then in a revised edition in 1963, and in *The Common Trench*,[60] a Research and Development Bulletin published by HMSO. The provision of services is considered in 'Sewers and public utility services', Section 7 of *Roads in Urban Areas*. Appendix 7, 'Provision for statutory undertakers' services where shared surfaces are used', of Design Bulletin 32 concerns the location of services in access ways (short lengths of culs-de-sac without footways, but where front gardens are provided).

CO-ORDINATION

The report of the Joint Committee on the Location of Underground Services recommends that the work of the various statutory undertakings be co-ordinated by the engineer of the appropriate highway authority. The statutory authorities should be approached as early as possible, preferably before any planning, to find out their requirements in general terms so that these can be taken into account from the earliest planning stage. In the case of new developments, once preliminary designs have been prepared, copies should be forwarded to the undertakers for them to show their requirements more precisely. This is the time to decide the final position of the services which best fulfils all the requirements of the particular design.

MATERIALS

Electricity   Cables are laid directly into the ground, except in particularly busy streets and across carriageways where they are drawn through 100mm diameter earthenware ducts (older 75mm ducts are still in use).
Gas   Mains are of cast iron or steel, 100mm minimum diameter. Service pipes are of steel, 25mm minimum diameter.
Water   Mains are mostly of spun or vertically cast iron and steel. They range in size between 50mm and 3.9m diameter but are commonly between 75mm and 300mm diameter with about half being 100mm diameter. Communication pipes are of lead or polythene, between 12mm and 50mm diameter.
GPO telephones   Polythene cables are laid in ducts or straight in the ground. Earthenware ducts in units of one, two, three, four, six and nine ways are the most popular. Asbestos cement ducts with 50mm and 80mm bore and 50mm pvc ducts are used on housing estates.

LOCATION UNDERGROUND

It is usual to locate services under footways or verges wherever possible rather than carriageways, (a) so that traffic will not be disrupted during emergency repairs, routine maintenance or provision of additional services;

and (b) because footways and verges over services can be constructed in materials that facilitate taking up and relaying and are less likely to suffer damage. The use of small scale units rather than in-situ paving also makes it easier to fit in access boxes and markers.

Sewers

Sewers cannot be grouped with the other services and generally have to be given priority of position as they form part of a far less flexible system. They have to be laid in straight lines and at uniform gradients between manholes (a manhole is necessary at each change of direction) and therefore often bear no relationship to local surface levels. They are commonly laid under the carriageway; properly constructed sewers rarely require repairs but with very wide roads or where it is foreseen that many branch connections will be required after the construction of the carriageway it is sometimes possible to duplicate the sewer if there is enough footway or verge space available.

Other services

Except in very low density developments the distribution of service mains is often duplicated so that connections do not have to be made across the carriageway. The Joint Committee on the Location of Underground Services has recommended a width of not less than 3.2m on both sides of a road, to allow distribution and service mains to be laid out in an orderly way, **B.36**, and allowing enough room for link disconnecting boxes and jointing pits in the electricity and telephone systems, and for valves and hydrants in the water system and for siphon pipes in the gas system. It also gives a reasonable space within which one utility can excavate and work without interfering with others.
If the footway is wider than 3.2m the arrangement of figure **B.37** should still be used in case of future road widening. Moderately sized services can be accommodated in a width of 2m, **B.36**.

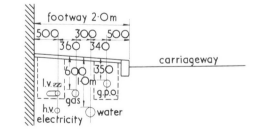

B.36 *Narrow footway*

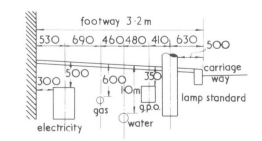

B.37 *Wide footway*

SPECIAL REQUIREMENTS

Electricity Low voltage cables should be at a minimum depth of 450 mm below the pavement surface. High voltage cables of 22 kV and over are subject to agreement with the highway authority. Underground link disconnecting boxes are required at intervals on the low voltage system, usually at street intersections with a pavement cover 760 × 600 mm. Runs of low voltage cables are usually restricted to 122 m lengths as it is wasteful on long runs. It consists of armoured cable with tile covers.

Gas Mains should be laid 600 mm to 760 mm deep; service pipes 450 mm to 600 mm. Access is rarely required to pipes throughout their lifetime. Access is usually required only to valves or to pumping pipes to remove condensate. Covers are usually 220 × 220 mm.

Water The cover required for mains and communication pipes is given in Table 24.

TABLE 24 Cover required for water mains and communication pipes

Size of main (m)	Depth of cover (m)
0.012–0.05	0.76
0.05–0.3	0.90
0.3–0.6	1.06
Over 0.6	1.20

Access is indeterminate and is required for repair and fixing new branches. Easily removable paving is desirable over water mains and at least 1.2 m diameter clear space above ground should be left around stopcocks.

GPO telephones Distance between jointing chambers should not exceed 155 m. The minimum depth for protected cable is 220 mm; for steel ducts 350 mm; and, for self-aligning ducts, 350 mm for one-way and 450 mm for multiple-way ducts when beneath a footway, and 600 mm in each case beneath a carriageway. Cover sizes range from 250 × 700 mm to 2.28 m × 700 mm.

CONCEALMENT OF COVERS

Manhole covers can be concealed by a layer of loose gravel or cobbles or by paving slabs on mortar dots. Markers must be provided to locate covers.

'THE COMMON TRENCH'

In *The Co-ordination of Underground Services on Building Sites, 1 The Common Trench*[60] a proposed cross section of a joint trench for housing estate distribution mains is presented. This cross section is reproduced here in figure **B.38** and Table 25.

TABLE 25 Proposed cross-section of joint trench for housing estate distribution mains (figure B.38)

Letter	Dimension		Remarks
	Imperial	Metric	
A	3′ 0″ min	900 mm min	
B	2′ 0″ min	600 mm min	
C	1′ 6″	450 mm	
D	6″ min	150 mm min	Without additional insulation
E	1′ 2″ min	350 mm min	
F	2′ 6″	750 mm	Distance of HV cable from GPO
H	1′ 0″ min	300 mm min	refers to multicore cable only.
J	6″ min	150 mm min	GPO require clearance of 1′ 6″
K	4″ min	100 mm min	from single core HV cable
L	8″ min	200 mm min	
M	9″ min	250 mm min	When MV included, otherwise 6″
N	5″ min	150 mm min	minimum (150 mm)
P	9″ min	250 mm min	
X	Variable	Variable	Gas pipe outside diameter
Y	Variable	Variable	Water pipe outside diameter
Z	Variable	Variable	HV cable outside diameter
R	= M + X + J		
S	= 2J + Y		
T	= J + N + Z		

Source: *HMSO (copyright reserved)*

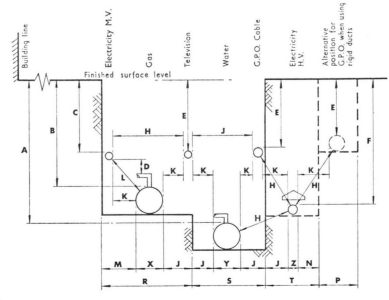

B.38 *Proposed cross-section of a joint trench for housing estate distribution*

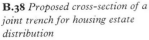

NOTES

1 C. Tunnard and B. Pushkarev, *Man-made America*, Yale University Press, New Haven, Conn., 1963

2 Ebenezer Howard, *Tomorrow: A Peaceful Path to Real Reform*, 1898

3 Paul Jacques Grillo, *Form, Function and Design*, Dover, New York, 1975

4 B. Adams and J. Noble, Housing: the Home in its Setting, *The Architects' Journal*, 11 September 1968

5 Thomas Sharp, *The Anatomy of the Village*, Penguin, Harmondsworth, 1946

6 Yoshinobu Ashihara, *Exterior Design in Architecture*, Van Nostrand, New York, 1970

7 Gordon Cullen, *Townscape*, Architectural Press, London, 1961

8 Department of the Environment, *Roads in Urban Areas*, HMSO, London, 1966

9 Department of the Environment, *The Layout of Roads in Rural Areas*, HMSO, London, 1968

10 J. F. Springer and K. E. Huizinga, *The road-picture as a touchstone for the three-dimensional design of roads*, Rijkswaterstaat Communications, Government Publishing Office, The Hague, 1975

11 *The Layout of Roads in Rural Areas, op. cit.*

12 American Association of State Highway Officials, *A Policy on Geometric Design of Rural Highways*, 1973

13 D. Appleyard, K. Lynch and J. R. Meyer, *The View from the Road*, MIT Press, Cambridge, Mass., 1964

14 J. R. Hamilton and Louis L. Thurstone, *Human Limitations in Automobile Driving*, 1937

15 Appleyard, Lynch and Meyer, *op. cit.*

16 American Association of State Highway Officials, *op. cit.*

17 L. J. Ritter and R. J. Paquette, *Highway Engineering*, The Ronald Press Co., 1967

18 Tunnard and Pushkarev, *op. cit.*

19 Sylvia Crowe, *The Landscape of Roads*, Architectural Press, London, 1960

20 J. E. Cordell and H. Howarth, *Principles and Practice of Road Location with particular reference to major roads outside built-up areas*, Institution of Civil Engineers, Road Engineering Division, London, 1947

21 K. B. Woods, *The Highway Engineering Handbook*, McGraw Hill, 1960

22 Tunnard and Pushkarev, *op. cit.*

23 American Association of State Highway Officials, *A policy on landscape development for the national system of interstate and defence highways*, 1961

24 K. B. Woods, *op. cit.*

25 Nan Fairbrother, *New Lives New Landscapes*, Architectural Press, London, 1972; Penguin, London, 1977

26 Michael Porter, *Some Aspects of Road Landscape*, seminar on Detailed Road Design, University of Sussex, June 1973

27 Warwickshire County Council, *Space Relations in Housing Areas*, 1973

28 Design Council, *Street Furniture from Design Index*, London, 1976

29 *ibid.*

30 Department of Transport, *Specification for Road and Bridge Works* (5th ed.), HMSO, London, 1976

31 Transport and Road Research Laboratory, *A Guide to the Structural Design of Pavements for New Roads*, Road Note 29, HMSO, London, 1970

32 L. W. Hatherley and P. C. Leaver, *Asphaltic Road Materials*, Edward Arnold, London, 1967

33 Transport and Road Research Laboratory and the Geological Survey and Museum, *Sources of Road Aggregate in Great Britain*, HMSO, London, 1968

34 Transport and Road Research Laboratory, *Sources of White and Coloured Aggregates in Great Britain*, Road Note 25, HMSO, London, 1966

35 Department of Transport, *Notes for Guidance on the Specification for Road and Bridge Works*, HMSO, London, 1976

36 Transport and Road Research Laboratory, *Recommendations for Road Surface Dressing*, Road Note 39, HMSO, London, 1972

37 *ibid.*

38 Transport and Road Research Laboratory, *Bituminous Materials in Road Construction*, HMSO, 1962

39 Transport and Road Research Laboratory, *The Design and Construction of Joints in Concrete Pavements*, Laboratory Report 512, HMSO, London, 1973

40 M. Gage and M. Vandenberg, *Hard Landscape in Concrete*, Architectural Press, London, 1975

41 M. Gage, *Guide to Exposed Concrete Finishes*, Architectural Press, London, 1970

42 H. S. Mildenhall, *Laying Precast Concrete Paving Flags*, Cement and Concrete Association, London, 1974

43 J. R. Collins and A. A. Lilley, *Laying Concrete Block Paving*, Cement and Concrete Association, London, 1976

44 A. J. Clark and A. A. Lilley, *Concrete Block Paving for Lightly Trafficked Roads and Paved Areas*, Cement and Concrete Association, London, 1978

45 A. A. Lilley and B. J. Walker, *Concrete Block Paving for Heavily Trafficked Roads and Paved Areas*, Cement and Concrete Association, London, 1978

46 Highway Research Board, American Association of State Highway Officials, *Road Test Reports 61A and 61E*, National Academy of Sciences, National Research Council, 1962

47 'Concrete Setts from Belgium', *Concrete Quarterly*, No 113, 1977

48 C. C. Handisyde, *Hard Landscape in Brick*, Architectural Press, London, 1976

49 L. Bovis and K. Thomas, *Brick Floors and Brick Paving*, The Brick Development Association Technical Note, Vol 1, No 7, May 1973

50 Transport and Road Research Laboratory, *Sources of White and Coloured Aggregates in Great Britain*, op. cit.

51 J. Gilchrist Wilson FRIBA, *White Concrete with Some Notes on Black Concrete*, Cement and Concrete Association, London, 1969

52 Elizabeth Beazley, *The Design and Detail of the Space between Buildings*, Architectural Press, London, 1960

53 Department of the Environment, *Roads in Urban Areas*, op. cit.

54 Department of the Environment, *The Layout of Roads in Rural Areas*, op. cit.

55 Department of the Environment, *Residential Roads and Footpaths: layout considerations*, Design Bulletin 32, HMSO, London, 1977

56 Department of the Environment, *Roads in Urban Areas*, op. cit.

57 *Access for Fire Appliances*, Fire Department Fire Prevention Note 1/70, Home Office, London, 1970

58 Department of the Environment, *Residential Roads and Footpaths*, op. cit.

59 Ann Thompson (ed.), *The Location of Underground Services*, The Institution of Civil Engineers, London, 1946; rev. ed. 1963

60 Ministry of Public Building and Works, *The co-ordination of underground services on building sites, 1 The Common Trench*, Research and Development Bulletin, HMSO, London, 1968

SELECT BIBLIOGRAPHY

PART I THE TOWNSCAPE ALIGNMENT

'Homes for tomorrow', *The Architects' Journal*, 9 May 1973

Ashihara, Yoshinobu, *Exterior Design in Architecture*, Van Nostrand, New York, 1970

Bacon, Edmund N., *Design of Cities*, Thames and Hudson, London, 1967

Beazley, Elizabeth, *Design and Detail of the Space Between Buildings*, Architectural Press, London, 1960

Burke, Gerald, *Towns in the Making*, Edward Arnold, London, 1975

Burke, Gerald, *Townscapes*, Penguin, Harmondsworth, 1976

Burt, M. E., *Roads and the Environment*, Transport and Road Research Laboratory Report 441, HMSO, London, 1972

City of Norwich Directorate of Planning and Environment, *Bowthorpe Design Guide*, 1975

Crawford, David, *A Decade of British Housing, 1963 to 1973*, Architectural Press, London, 1975

Cullen, Gordon, *Townscape*, Architectural Press, London, 1961

Department of the Environment, 'The design of streets and other spaces', Area Improvement Note 6, HMSO, London, 1973

Department of the Environment, 'The estate outside the dwelling; reaction of residents to aspects of housing layout', Design Bulletin 25, HMSO, London, 1972

Department of the Environment, 'Landscape of new housing', Housing Development Notes, Housing Development Directorate, 1974

Department of the Environment, 'Landscaping for flats', Design Bulletin 5, HMSO, London, 1967

Department of the Environment, 'Local authority housing: a comparative study of the land use and built form of 110 schemes', Occasional Paper 2/73, Housing Development Directorate, August 1973

Department of the Environment 'The quality of local authority housing schemes', Occasional Paper 1/74, Housing Development Directorate, 1974

Eckbo, Garratt, *Urban Landscape Design*, McGraw Hill, Maidenhead, 1974

Essex County Council, *A Design Guide for Residential Areas*, 1973

Gibberd, Frederick, *Town Design*, Architectural Press, London, (5th ed.) 1967

Greater London Council, *Greater London Development Plan*, July 1976

Grillo, Paul Jacques, *Form, Function and Design*, Dover, New York, 1975

Halprin, Lawrence, *Cities*, Reinhold, New York, 1963

Hesselgren, Sven, *The Language of Architecture*, Applied Science Publishers, London, 1969

Howard, Ebenezer, *Tomorrow: A Peaceful Path to Real Reform*, 1898

Isaac, A. R. G., *Approach to Architectural Design*, Iliffe, 1971

Jacobs, Jane, *The Death and Life of Great American Cities*, Penguin, Harmondsworth, 1974

Licklider, H., *Architectural Scale*, Architectural Press, London, 1965

Lynch, Kevin, *The Image of the City*, MIT Press,

Cambridge, Mass., 1975

Lynch, Kevin, *Site Planning*, MIT Press, Cambridge, Mass., 1975

Ministry of Housing and Local Government, *Homes for Today and Tomorrow* (The Parker Morris Report), HMSO, London, 1961

Mumford, Lewis, *The City in History*, Secker and Warburg, London, 1961; Penguin, Harmondsworth, 1961

Nairn, Ian, *Counter Attack*, Architectural Press, London, 1957

Nairn, Ian, *Outrage*, Architectural Press, London, 1955

Newman, Oscar, *Defensible Space : people and design in the violent city*, Architectural Press, London, 1973

Noble, J. and Adams, B., 'Housing: the home in its setting', *The Architects' Journal*, 11 September 1968

Norberg- Schultz, Christian, *Existence, Space and Architecture*, Praeger, New York, 1971

Norberg-Schultz, Christian, *Meaning in Western Architecture*, Studio Vista, London, 1975

'The organisation of space in housing neighbourhoods', Report of a symposium held at the Royal Institute of British Architects, Institute of Landscape Architects, London, 1961

Purdon, C. B., *The Building of Satellite Towns*, Dent, London, 1949

Ritter, Paul, *Planning for Man and Motor*, Pergamon, Oxford, 1964

Rowland, Kurt, *The Shape of Towns* (Looking and Seeing, Number 4), Ginn and Co., Lexington, Mass., 1966

Segal, Walter, *Home and Environment*, Leonard Hill, London, 1953

Sharp, Thomas, *The Anatomy of the Village*, Penguin, Harmondsworth, 1946

Sharp, T., Gibberd, F. and Holford, W. G., *Design in Town and Village*, Ministry of Housing and Local Government, HMSO, 1953

Sitte, Camillo, *The Art of Building Cities*, Vienna, 1889

Sitte, Camillo, *City Planning according to Artistic Principles*, trs. George and Christiane Collins, Phaidon, Oxford, 1965

Smith, Peter, *The Dynamics of Urbanism*, Hutchinson Educational, London, 1974

Spreiregen, P. D., *The Architecture of Towns and Cities*, McGraw-Hill, Maidenhead, 1965

Stein, Clarence, *Towards New Towns for America*, MIT Press, Cambridge, Mass., 1966

Stevens, P. S., *Patterns in Nature*, Penguin, Harmondsworth, 1976

Tandy, C. (ed.), *Handbook of Urban Design*, Architectural Press, London, 1972

Thompson, F. L., *Site Planning in Practice*, Oxford Technical Publications, 1923

Tuan, Yi-Fu, *Topophilia*, Prentice-Hall, Englewood Cliffs, N.J., 1974

Warwickshire County Council, *Space Relations in Housing Areas*, County Planning Department, 1973

Weddle, A. E. (ed.), *Techniques of Landscape Architecture*, Heinemann, London, 1967

de Wolfe, Ivor, *Civilia: the end of sub urban man*, Architectural Press, London, 1971

de Wolfe, Ivor, *Italian Townscape*, Architectural Press, London, 1963

de Wolfe, Ivor, 'Sociable Housing', *The Architectural Review*, October 1973

Worskett, R., *The Character of Towns*, Architectural Press, London, 1969

PART II THE FLOWING ALIGNMENT

Allan, B. J., 'Aesthetics in highway engineering', *Journal of the Institute of Highway Engineers*, 1968

American Association of State Highway Officials, *A Policy on Geometric Design of Rural Highways*, 1973

Appleyard, D., Lynch, K. and Meyer, J. R., *The View from the Road*, MIT Press, Cambridge, Mass., 1964

Christy, D. A. T., 'Construction, landscaping and the environment', *Civil Engineering and Public Works Review*, June 1970

Crowe, S., *The Landscape of Roads*, Architectural Press, London, 1960

The Council for the Preservation of Rural England, *The Landscape Treatment of Roads*, London, 1954

Council for the Protection of Rural England, *Roads in the Landscape*, London, 1971

Department of the Environment, *Layout of Roads in Rural Areas*, HMSO, London, 1968

Department of the Environment, *Roads in Urban Areas*, HMSO, London, 1966

Halprin, L., *Freeways*, Reinhold, New York, 1966

Hamilton, J. R. and Thurstone, L. L., *Human Limitations in Automobile Driving*, 1937

'Highway location as a problem of urban and landscape design', *Highway Research Record*, 1963

Jellicoe, G. A., 'Studies in landscape design', *Motorways*, 1960

'Landscaping of Motorways', a conference at the Institution of Civil Engineers, 22 May 1962, The British Road Federation, 1963

Ministry of Transport, *Roads in the Landscape*, HMSO, London, 1967

Porter, M. R., 'Motorways in the rural environment', *Journal of the Institute of Highway Engineers*, 1970

'Roads in the landscape', international conference organized by the Ministry of Transport and the British Road Federation, University of Keele, 1967

Robinson, John, *Highways and our Environment*, McGraw-Hill, Maidenhead, 1971

Snow, W. B., *The Highway and the Landscape*, Rutgers University Press, New Brunswick, N.J., 1959

Spearing, G. D. and Porter, M. R., 'Motorways and the rural environment', Proceedings of Institution of Civil Engineers, October 1971

Spencer, W. H., *The Co-ordination of Horizontal and Vertical Curves in the Alignment of High-Speed Roads*, The Institute of Civil Engineers, 1949

Springer, J. F. and Huizinga, K. E., *The Road-picture as a Touchstone for the Three-dimensional Design of Roads*, Rijkswaterstaat Communications, Government Publishing Office, The Hague, 1975

Tunnard, C. and Pushkarev, B., *Man-made America*, Yale University Press, New Haven, Conn., 1963

US Department of Commerce, *A Proposed Program for Scenic Roads and Parkways*, US Government Printing Office, 1965

Van Riper, D., 'Aesthetics in highway design', *Civil*

Engineering, April 1972

Varming, M., *Motorveje Landskabet*, Staten Bygge-forskningsinstitut, Copenhagen, 1970

Yamanaka, H. M., 'Highway cross section combines aesthetics with safety', *Public Works*, February 1970

Woods, K. B., *The Highway Engineering Handbook*, McGraw-Hill, Maidenhead, 1960

PART III   LANDFORM AND PLANTING

The American Association of State Highway Officials, *A Policy on Landscape Development for the National System of Interstate and Defence Highways*, 1961

Anderson, M. L., *The Selection of Tree Species: an ecological basis of site classification for conditions found in Great Britain and Ireland*, Oliver and Boyd, Edinburgh, 1961

Arboricultural Association, *A Guide to Tree Pruning*, advisory leaflet No. 2, 1970

Arboricultural Association, *A Tree for Every Site*: a select list of trees and shrubs, leaflet No. 5, H. G. Hillier, 1971

Arboricultural Association, *Tree Preservation Orders*, Advisory leaflet No. 1, 1967

Bernatsky, A., 'Trees in the city', *Garten und Landschaft*, Munich, October, November and December, 1974

Boddy, F. A., 'Street tree planting with a difference', *Parks and Recreation*, November 1974

Boddy, F. A., *Highway Trees*, Clarke and Hunter, Guildford, 1968

Colvin, Brenda, *Trees for Town and Country*, Lund Humphries, London, 1972

Cordell, J. E. and Howarth, H., *Principles and Practice of Road Location with particular reference to major roads outside built-up areas*, Institution of Civil Engineers, Road Engineering Division, 1947

Correy, A., 'Trees in streets rethought', *Architecture in Australia*, October 1972

Council for the Protection of Rural England, *Roads in the Landscape*, 1971

Crowe, Sylvia, *Tomorrow's Landscape*, Architectural Press, London, 1956

Department of the Environment, *Grass cutting and hedgerow treatment on trunk roads and motorways*, Technical Memorandum No H 9/75, Highways Planning and Management, 1975

Department of the Environment, *Landscaping for Flats*, Design Bulletin 5, Housing Development Directorate, HMSO, London, 1969

Department of the Environment, *Landscape of New Housing*, Part 1 Background (February 1973) Part 2 Trees in housing (February 1973) Part 3 Shrubs in housing (February 1974) Part 4 Grass and other small plants (February 1974), Housing Development Notes, Housing Development Directorate

Design Council, *Street furniture from Design Index*, London, 1976

Dundee Corporation, *Landscaping in the Urban Environment*, Parks and Recreation Department, October 1971

Dyke, J., 'Landscape of motorways', *Civil Engineering and Public Works Review*, June 1970

Edwards, P., *Trees and the English Landscape*, Bell, London, 1962

'Evaluation of the types of trees to be used in urban street

areas', *Garten und Landschaft*, Munich, December 1976

Fairbrother Nan, *New Lives, New Landscapes*, Architectural Press, London, 1972; Penguin, Harmondsworth, 1977

Flemer, W., 'Trees in towns', *Royal Horticultural Society Journal*, January 1975

Grigson, R. R. W., 'Preserving trees during road works', *Surveyor*, 24 July, 1970

Hicks, P., *The Care of Trees on Development Sites*, Advisory leaflet No. 3, Arboricultural Association, 1970

Hilliers' Manual of Trees and Shrubs, Hillier and Sons, 4th ed., Winchester, 1974

James, N. D. G., *The Arboriculturist's Companion*, Blackwell, Oxford, 1972

Jellicoe, G. A., 'Motorways, their landscaping, design and appearance', paper to the Town Planning Institute, 1958

Luz, H., 'Bringing plants into towns', *Garten und Landschaft*, Munich, June 1973

McHarg, I. L., *Design with Nature*, Natural History Press, 1969

Ministry of Housing and Local Government, *Trees in Town and City*, HMSO, London, 1958, reprinted 1967

Morling, Ronald J., *Trees: including preservation, planting, law and highways*, Gazette Press, 1963

Patterson, G., 'Trees in urban areas (design with plants)', *Landscape Design*, November 1975

Porter, Michael, 'Some aspects of road landscape', seminar on Detailed Road Design, University of Sussex, June 1973

Property Services Agency, *Design with Trees*: Part 1 Existing trees and buildings, Part 2 Tree work, HMSO, London, 1974

Property Services Agency, *Design with Trees*: planting, HMSO, London, 1974

Radd, A., 'Trees in towns and their evaluation', *Arboricultural Association Journal*, September 1976

Ritter, L. J. and Paquette, R. J., *Highway Engineering*, The Ronald Press, 1967

Snow, W. B. (ed.), *The Highway and the Landscape*, Rutgers University Press, New Brunswick, N.J., 1959

Thomas, Graham Stuart, *Plants for Ground Cover*, Dent, London, 1970

Warwickshire County Council, *Space Relations in Housing Areas*, 1973

Wells, D. V., *Trees—site preparation and planting*, Advisory leaflet No. 4, Arboricultural Association, 1970

West Sussex County Council, *Tree Planting in Urban Areas*, Chichester, 1972

Woods, K. B., *The Highway Engineering Handbook*, McGraw-Hill, Maidenhead, 1960

Relevant British Standards and Codes of Practice
CP 2003:1959 Earthworks
BS 2468:1963 Glossary of terms relating to agricultural machinery and implements
BS 3882:1965 Recommendations and classification for top soil
BS 3936 Nursery stock
Part 1: 1965 Trees and shrubs
Part 4: 1966 Forest trees
BS 3969:1965 Recommendations for turf for general landscape purposes
BS 3975 Glossary for landscape work

Part 4: 1966 Plant description
Part 5: 1969 Horticultural, arboricultural and forestry practice
BS 3998:1966 Recommendations for tree work
BS 4043:1966 Recommendations for transplanting semi-mature trees
BS 5236:1975 Recommendations for the cultivation and planting of trees in the extra large nursery stock category

PART IV MATERIALS, SURFACE AND TRIM

Asphaltic construction: general references

Asphalt and Coated Macadam Association, *Hot Process Asphalts*, London, 1975

Asphalt and Coated Macadam Association, *Model Specification for Roads and Footways on Housing Estates*, London, 1975

Asphalt and Coated Macadam Association, *Modern Flexible Road Construction*, London, 1976

Asphalt and Coated Macadam Association, *The Types and Scope of Coated Macadam*, London, 1975

Department of Transport, *Specification for Road and Bridge Works*, HMSO, London, 1976

Department of Transport, *Notes for Guidance on the Specification for Road and Bridge Works*, HMSO, London, 1976

Department of Transport, *Road pavement design*, Technical Memorandum No H6/78, HMSO, London, 1978

Department of Transport, *Supplement No 1 to the Specification for Road and Bridge Works and the Notes for Guidance on the Specification for Road and Bridge Works*, HMSO, London, 1978

Hatherly, L. W. and Leaver, P. C., *Asphaltic Road Materials*, Edward Arnold, London, 1967

Krebs, R. D. and Walker, R. D., *Highway materials*, McGraw-Hill, Maidenhead, 1971

Transport and Road Research Laboratory, *A Guide to the Structural Design of Pavements for New Roads*, Road Note 29, HMSO, London, 1976

Transport and Road Research Laboratory, *Bituminous Materials in Road Construction*, HMSO, London, 1962

Transport and Road Research Laboratory, *Recommendations for Road Surface Dressing*, Road Note 39, HMSO, London, 1972

Transport and Road Research Laboratory and the Geological Survey and Museum, *Sources of Road Aggregate in Great Britain*, HMSO, London, 1968

Transport and Road Research Laboratory, *Sources of White and Coloured Aggregates in Great Britain*, Road Note 25, HMSO, London, 1966

Wallace, H. A. and Martin, J. R., Asphaltic Pavement Engineering, McGraw-Hill, Maidenhead, 1967

Asphaltic construction: relevant British Standards
BS 76:1974   Tars for road purposes
BS 434:   Bitumen road emulsions (anionic and carionic)
              Part 1: 1973 Requirements
              Part 2: 1973 Recommendations for use
BS 594:1973   Rolled asphalt (hot process) for roads and other paved areas
BS 812:   Methods for sampling and testing of mineral aggregates, sands and fillers
BS 1446:1973 Mastic asphalt (natural rock asphalt fine aggregate) for roads and footways
BS 1447:1973 Mastic asphalt (limestone fine aggregate) for roads and footways
BS 1984:1967 Gravel aggregates for surface treatment (including surface dressings) on roads
BS 3690:1970 Bitumens for road purposes
BS 4987:1973 Coated macadam for roads and other paved areas
BS 5273:1975 Dense tar surfacing for roads and other paved areas

Concrete construction: general references
American Association of State Highway Officials Road Test, *Reports 61A and 61E*, Highway Research Board, National Academy of Sciences, National Research Council, 1962

Blake, L. S., *Recommendations for the Production of High Quality Concrete Surfaces*, 1967

British Precast Concrete Federation, *Kerbs, Channels, Edgings and Quadrants*, 1967

—*Pavement Lights*, 1969

—*Fencing* (British Standard types), 1969

Cement and Concrete Association, *Concrete in Print*, January 1978, a catalogue of textbooks, proceedings, technical reports data sheets, translations, information booklets available from the Cement and Concrete Association and including the following: *Concrete as a visual material*, P. Marsh, 1974; *Concrete block paving*, leaflet INF 2/78, 1978; *Concrete block paving for heavily trafficked roads and paved areas*, 1978; *Concrete block paving for lightly trafficked roads and paved areas*, 1978; *Decorative paving patterns in concrete*, 1962; *Precast concrete production*, J. G. Richardson, 1973; *Recommendations for the production of high quality concrete surfaces*, L. O. Blake, 1967; *A specification for concrete block paving*, 1978; *Specification clauses covering the production of high quality surface finishes*, J. Gilchrist Wilson, 1970; *White concrete with some notes on black concrete*, J. Gilchrist Wilson, 1969.

Cement and Concrete Association, Films and Photographs Department, *Cement Block Paving*, a 16mm film in colour with sound, running time 14mins, available free on loan in UK

Cement and Concrete Association, *Report on a Study Visit to Belgium and the Netherlands*, Departmental Note DN/2044 (Confidential Report), 1974

Cement and Concrete Association, *A Study Visit to Germany and Denmark*, Departmental Note DN/2046 (Confidential Report), 1975

Concrete Quarterly, 'Concrete Setts from Belgium', No 113, 1977

Gage, M., *Guide to Exposed Concrete Finishes*, Architectural Press, London, 1975

Gage, M. and Vandenberg, M., *Hard Landscape in Concrete*, Architectural Press, London, 1975

Lilley A. A. and Collins, J. R., *Laying Concrete Block Paving*, Cement and Concrete Association, London, 1976

Mildenhall, H. S., *Laying Precast Concrete Paving Slabs*, Cement and Concrete Association, London, 1974

Sharp, D. R., *Concrete in Highway Engineering*, Pergamon Press, Oxford 1970

Transport and Road Research Laboratory, *Concrete Roads*, HMSO, London, 1955

Transport and Road Research Laboratory, *Sources of White and Coloured Aggregates in Great Britain*, Road Note 25, HMSO, London, 1966

Transport and Road Research Laboratory and The Geological Survey and Museum, *Sources of Road Aggregates in Great Britain*, HMSO, London, 1968

Walker, J. C., *The Performance of Concrete Roads and Motorways during the Past Decade*, Concrete Society, London, 1967

Concrete construction : relevant British Standards and Codes of Practice

CEMENT

BS 890:1972   Building limes

BS 12:   Portland cement (ordinary and rapid-hardening)
Part 1: 1958 Imperial units
Part 2: 1971 Metric units

BS 146:   Portland-blastfurnace cement
Part 1: 1958 Imperial units
Part 2: 1973 Metric units

BS 1370:   Low heat Portland cement
Part 2: 1974 Metric units

BS 4027:   Part 2: 1972 Metric units, sulphate-resisting Portland cement

BS 4246:   Part 2: 1974 Metric units, low heat Portland-blastfurnace cement

BS 4248:1974   Supersulphated cement

BS 4550:   Methods of testing cement
Part 2: 1970 Chemical tests

BS 4627:1970   Glossary of terms relating to types of cements, their properties and components

AGGREGATE

BS 63:   Single-sized roadstone and chippings
Part 1: 1951 Imperial units
Part 2: 1971 Metric units

BS 812:   Methods for the sampling and testing of mineral aggregates, sands and fillers

BS 882, 1201:   Part 1: 1965 Aggregates from natural sources for concrete (including granolithic)
Part 2: 1973 Metric units

BS 1047:   Part 2: 1974 Metric units, air-cooled blast-furnace slag coarse aggregate for concrete

BS 1165: 1966 Clinker aggregate for concrete

BS 1198:1200:1976 Building sands from natural sources

BS 1984:1967 Gravel aggregates for surface treatment (including surface dressings) on roads

BS 2451:1963 Chilled iron shot and grit

CONCRETE

BS 1926:1962 Ready-mixed concrete

BS 1881:   Methods of testing concrete
Parts 1–5
Part 6: 1971 Analysis of hardened concrete

BS 4408:   Non-destructive methods of test for concrete, Parts 1–5

MISCELLANEOUS

BS 340:1963   Pre-cast concrete kerbs, channels, edgings and quadrants

BS 368:1971   Precast concrete flags

BS 556:   Concrete cylindrical pipes and fittings including manholes, inspection chambers and street gullies
Part 1: 1966 Imperial units
Part 2: 1972 Metric units

BS 1014:1975   Pigments for Portland cement and Portland cement products

BS 1308:1970   Concrete street lighting columns

BS 2499:1973   Hot applied joint sealants for concrete pavements

BS 5212:1975   Cold poured joint sealants for concrete pavements

CODES OF PRACTICE

CP 110   The structural use of concrete

GLOSSARIES

BS 2787:   Glossary for concrete and reinforced concrete

BS 3683:   Glossary of terms used in non-destructive testing
Parts 1 and 2: 1963
Parts 3: 1964
Parts 4 and 5: 1965

BS 4340:1968 Glossary of formwork terms

Other references : general

Beazley, Elizabeth, *Design and Detail of the Space between Buildings*, Architectural Press, London, 1960

Brick Development Association, *The Brick Bulletin*, and the following publications:
—*Brick Floors and Brick Paving*, Technical Note Volume 1, No 7, May 1973
—*Cleaning of Brickwork*, Practical Note 4, May 1974
—*Mortars for Brickwork*, Practical Note 2, September 1973

Brick Institute of America, *Brick Floors and Pavements*, Technical Note, 1975

Building Research Establishment, *Mortars for Brickwork*, Digest 160 (second series) 1973
—*Sulphate Attack on Brickwork*, Digest 89 (second series), 1971

Design Council, *Street Furniture from Design Index*, London, 1976

Handisyde, C. C., *Hard Landscape in Brick*, Architectural Press, London, 1976

Lovejoy, Derek and Partners, *Spon's Landscape Handbook*, E. and F. N. Spon Ltd., London, 1975–6

Lovejoy, Derek and Partners, *Spon's Landscape Price Book*, E. and F. N. Spon Ltd., London, 1978

Scottish Local Authorities' Special Housing Group Research Unit, *Hard Surfaces : Edging*, Edinburgh, 1975

Transport and Road Research Laboratory Reports (HMSO, London):
154 *Estimates of the proportion of space occupied by roads and footpaths in towns*, 1968
291 *The development of rumble areas as a driver alerting device*, 1977

409 *The influence of visual pattern on perceived speed,* 1971

Relevant British Standards and Codes of Practice

BS 187:   Part 2: 1970 Calcium silicate (sandlime and flintlime) bricks, metric units (plus Amendment AMD 695, February 1971)

BS 435:1975 Dressed natural stone kerbs, channels, quadrants and setts

BS 497   Specification for manhole covers, road gully gratings and frames for drainage purposes
             Part 1: 1976 Cast iron and cast steel

BS 873   The construction of road traffic signs and internally illuminated bollards, Parts 1–4

BS 892:1967 Glossary of highway engineering terms

BS 1377:1975 Methods of test for soil for civil engineering purposes

BS 1722   Fences
   Part 1:1972   Chain link fences
   Part 2:1973   Woven wire fences
   Part 3:1973   Strained wire fences
   Part 4:1972   Cleft chestnut pale fences
   Part 5:1972   Close-boarded fences including oak pale fences
   Part 6:1972   Wooden palisade fences
   Part 7:1972   Wooden post and rail fences
   Part 8:1966   Mild steel or wrought iron continuous bar fences
   Part 9:1963   Mild steel or wrought iron unclimbable fences with round or square verticals and flat standards and horizontals
   Part 10:1972 Anti-intruder chain link fences
   Part 11:1972 Woven wood fences

BS 1840:1960 Steel columns for street lighting
BS 3921:1974 Clay bricks and blocks
BS 3989:1966 Aluminium street lighting columns
BS 4729:1971 Shapes and dimensions of special bricks
CP 1004   Road lighting, Parts 1–9

Design Data

General references
Blake, L. S., *Civil Engineers' Reference Book*, Newnes-Butterworth, Borough Green, Kent, 1975
British Road Federation, 'Basic road statistics', *The Federation* (annual)
Cheshire County Council, *Housing: roads*, Design Aid, 1976
Department of the Environment, *Cars in Housing 2*, Design Bulletin 12, HMSO, London, 1971

—*Housing: needs and action*, Circular 24/75, HMSO, London, 1975
—*Parking in New Housing Schemes*, Parts 1 and 2, Housing Development Note 7, HMSO, London, 1977
—*Residential Access Roads and Footpaths*, Letter circulated to all Highways, Planning and Housing Officers, June 1975
—*Residential Roads and Footpaths: layout considerations*, Design Bulletin 32, HMSO, London, 1977
—*Traffic in General Improvement Areas*, Area Improvement Note No 9, HMSO, London, 1974
Duffell, J. and Hopper, R., 'Sleeping policemen: their effectiveness in regulating vehicle speeds', *Municipal Engineers' Journal*, August 1975
Emmerson, J., 'A note on speed-road curvature relationships', *Traffic Engineering and Control*, Vol 12, November 1970
Emmerson, J., 'Speed of cars on sharp horizontal curves', *Traffic Engineering and Control*, Vol 12, July 1969
Fire Department, *Access for Fire Appliances*, Fire Prevention Note 1/70, Home Office, London, 1970
Ministry of Public Building and Works, *Co-ordination of underground services on building sites, 1 The Common Trench*, Research and Development Bulletin, HMSO, London, 1968
'New registrations of new motor vehicles in Great Britain, Northern Ireland and the Isle of Man', *SMMT* (six monthly), Society of Motor Manufacturers and Traders Ltd.
'Specifications and prices of British and foreign cars', *Motor Transport Press* (annual)
Sando, E. D. and Batty, V., *Road traffic and the Environment*, Social trends No 5, Central Statistical Office, 1974
Thompson, Ann (ed.), *The location of Underground Services*, The Institution of Civil Engineers, 1946; rev. ed. 1963
Transport and Road Research Laboratory Reports (HMSO, London):

   71 *Single track roads in the Scottish Highlands* (further traffic studies, 1964), M. S. Walker, J. W. Tyler and J. R. Lake, 1967
   88 *The relation between a driver's speed and his accident rate*, J. M. Munden, 1967
   97 *The effect of speed and speed change on drivers' speed judgement*, G. G. Denton, 1967
   608 *Road width requirements of commercial vehicles when cornering*, G. Brock, 1973
   668 *The Hampshire child pedestrian accident study*, G. B. Grayson, 1975

Codes of practice
CP 306 Part 1: 1972, The storage and collection of refuse for residential buildings
CP 2001:1957 Site investigations

CREDITS

Grateful acknowledgement is made to the following for permission to reproduce the material listed (numbers refer to figures):

Calder and Boyars Limited: quotation on p. 11; Greater London Council, Department of Architecture and Civic Design: 1.4, 1.10, 1.14, 1.29, 1.30, 1.34, 1.35, 1.36, 1.39, 1.40, 1.42, 1.43, 1.44, 1.47, 1.48, 1.49, 1.50, 1.51, 1.52, 1.55, 1.56, 1.63, 1.64, 1.65, 1.66; extract on p. 34; 1.67, 1.68, 1.70, 2.59, 2.63, 3.5, 8.101, 8.102, 9.31, 10.5, 10.18, 10.19, 10.24, 10.29, 10.33, 11.12; Aerofilms Limited: 1.12, 3.23, 3.45; J. M. Dent & Sons (*The Building of Satellite Towns*, C. B. Purdon, 1949) and the executors for Louis de Soissons: 1.20, 2.64; Milton Keynes Development Corporation, Department of Architecture: 1.23, 1.41, 1.69; Chapman Taylor Partners: 1.28; Basildon Development Corporation, Department of Architecture: 1.32, 1.33, 1.54; Whicheloe Macfarlane Partnership and Building Partnership (Bristol) Ltd: 1.37; Architectural Press: 1.38 and 9.19–9.27 (*Hard Landscape in Concrete* by M. Gage and M. Vandenberg, 1975), 9.12–9.18, 9.54 and 9.55, Table 5 on p. 237, Table 8 on p. 251 and Table 9 on p. 251 (*The Architects' Journal*); Runcorn Development Corporation, Department of Architecture: 1.45, 1.46, 1.59, 2.53; Mr E. Jenkins, Highway Engineering Department, Runcorn Development Corporation: 1.53; Telford Development Corporation, Department of Architecture: 1.58; John Spence and Partners, Architects and Crown Estate Commissioners, Developers: 1.60, 1.61, 1.62; Faber and Faber Limited and Keith Gibson (photographer): 2.1, 2.2 (*The Landscape of Ideas*, P. Nutgens, 1972); Warrington Development Corporation, Department of Architecture: 2.50, 2.58, 2.70; J. Whittle and Partners, Architects: 2.52; Cheshire County Council, County Planning Department: 2.69, Director of Highways and Transportation: 2.74; *Building*: 2.118; Mr E. Watson, Chief Architect and Planning Officer, Washington Development Corporation: 2.121, 2.122; Andrews Sherlock and Partners: 2.129; Essex County Council, Planning Department: 2.132, 3.18; City of Norwich, Department of Planning and Environment: 2.133; Penguin Books and the executors for Thomas Sharp: 2.142, 2.144, 2.146, 2.148, 2.150, 2.153, 2.155, 3.47 (*The Anatomy of the Village*, Thomas Sharp); Applied Science Publishers Limited (*The Language of Architecture*, Sven Hesselgren, 1969): 3.36, 3.45; Yale University Press: extract on p. 133, Cement and Concrete Association: 9.30, 9.40–9.44, 9.45–9.47, Table 7 on p. 246, 10.7, 10.20; West German Road Research Association: 5.35; MIT Press, Cambridge, Massachusetts, for permission to base the following sketches on illustrations appearing in their publication *The View from the Road* by D. Appleyard, K. Lynch and J. R. Meyer (1964): 6.2–6.4, 6.7, 6.11 and 6.12; Rijkswaterstaat Communications, Government Publishing Office, The Hague, for permission to base the following sketches on illustrations appearing in their publication *The Road-picture as a Touchstone for the Three-Dimensional Design of Roads* by J. F. Springer and K. E. Huizinga (1975): 5.14–5.16, 5.19, 5.22, 5.37–5.39, 5.41, 5.43–5.46, 5.47, 5.49, 5.56; The Institution of Civil Engineers, for permission to base sketches 5.53b and 5.54b on a diagram in *The Co-ordination of Horizontal and Vertical Curves in the Alignment of High Speed Roads* by W. H. Spencer (1949); Transport and Road Research Laboratory: 5.10,

5.30, 8.4, Table 2 on p. 227 and Table 3 on p. 228, 9.8; The National Gallery, London: 8.27; Warwickshire County Council, Planning Department: 8.59, 8.60; Design Council: quotation on p. 204; 8.85 (*Street Scene*, 1976); Property Services Agency (*Design with Trees*, 1974): 8.88–8.90; British Standards Institution: extract on pp. 219 ff; The Controller of Her Majesty's Stationery Office: Table 2 on p. 227, Table 3 on p. 228, 9.8, Table 18 on p. 287; Charcon Products Limited: 9.29; Mono Concrete Limited: 9.32, 9.33, 9.34, 9.37, 9.38, 9.50; Grass Concrete Limited: 9.35, 9.36; Brooklyns Westbrick Limited: 9.39; *Concrete Quarterly* (Editor, George Perkin): 9.48, 9.49; British Engineering Brick Association: 9.53.

Material appearing in Appendix B, section 2 Local Distributors, is reproduced by kind permission of the Controller of Her Majesty's Stationery Office, as follows: From *Roads in Urban Areas* (1966): tables 14 and 15; data on lane widths for distributor roads, road radii and widening at bends, maximum and minimum crossfalls; figure B. 2 and Table 17; data on footway widths; figure B.3 and data on a typical cross-section; data on junction spacing. From *Residential Roads and Footpaths: layout considerations*, Design Bulletin 32, (1977): figures B.6–B.16 and related data; data on stopping distances and parking distances; figures B.18, B.19 and B.21, B.22–29 and accompanying data; data on gradients; figures B.30–35. From *The Co-ordination of Underground Services on Building Sites* (1968): figure B.39 and Table 18. From *Technical Memorandum* H9/76: tables 12 and 13; data on visibility at junctions and corner radii; figures B.36 and B.37.

Location of photographs

The location of photographs is given below, with the exception of those cases where it has already been given in the text:

INDEX